The DSMC method

G. A. Bird

Emeritus Professor, The University of Sydney

Version 1.2, 2013

ISBN 9781492112907

Graeme Austin Bird

B.Sc. (1951), B.E. (1953), M.E. (1959), and Ph.D. (1963),
  all from the University of Sydney.

Foreign Associate, National Academy of Engineering of the United States
Fellow, Australian Academy of Technological Sciences and Engineering.
Fellow, American Institute of Aeronautics and Astronautics.
Fellow, Royal Aeronautical Society.
Fellow, Institution of Engineers, Australia.

NASA Medal for Exceptional Scientific Achievement, 1988.
AIAA Thermophysics Award, 1988.

Permanent Appointments:
  1953 --59 Australian Defence Scientific Service, Scientific Officer.
  1960 --   University of Sydney
    1960--1964 Lecturer, Senior Lecturer in Aeronautical Engineering
    1964--1990 Lawrence Hargrave Professor of Aeronautical Engineering
    1990 --   Professor Emeritus
  1982 --   Managing Director, GAB Consulting Pty. Ltd.

Temporary Appointments (6 months or more):
  1955   College of Aeronautics, Cranfield, U.K.   Attached Scientist.
  1956   RAE, Farnborough, U.K. Attached Scientist.
  1964   University of Manchester, Simon Research Fellow.
  1969   California Institute of Technology, Visiting Professor.

Temporary Appointments (less than 6 months):
  1974   Imperial College, London.
  1975   Old Dominion University (NASA Langley)
  1976   Max-Planck-Institut-fur-Stromungsforschung, Göttingen.
  1998   Kyoto University
  2012 & 2013   Hong Kong University of Science and Technology

PREFACE

More than fifty years have now elapsed since the writing of the first paper on the Direct Simulation Monte Carlo, or DSMC, method. The early DSMC procedures were justifiably regarded with some suspicion and, by modern standards, computers were extremely slow and expensive. The early developments were described in the first monograph written by the author. The title was *Molecular Gas Dynamics* and it was published by Oxford University Press in 1976. Given the general lack of acceptance of the DSMC method at that time, the DSMC method was presented in a somewhat tentative manner and appeared only in the later chapters. By the time that the second monograph *Molecular Gas Dynamics and the Direct Simulation of Gas Flows* appeared in 1994, the DSMC procedures were sounder and more comprehensive, computers were incomparably faster and, most importantly, the method had become widely accepted. At the same time, the description of the DSMC method was again confined to the latter half of the book. The background material associated with the words "molecular gas dynamics" has not changed greatly over the past two decades, but almost all the DSMC procedures have been either revised or superseded. The writing of this book has been motivated by the need to document these developments.

The background material has been reduced to cover only the results that are necessary to support the description of the DSMC procedures. The 1994 monograph was concerned with alternative methods and the derivation of the background equations. This affected the order of presentation of the key results that are required for DSMC. This book is not concerned with the derivation of standard equations and the priority is to present them in a form that makes them easily accessible to the user. While this has simplified the task of writing the book, the recent proliferation of DSMC papers that present alternative procedures makes it difficult to assess them all in order to determine the optimum set of procedures. This book is largely based on the procedures that are currently employed by the author, but they are under constant review and will be changed when alternative procedures are found to be preferable.

The DSMC results that are presented in this book have been made either with the open source computer code **DSMC** that forms part of the book or with the **DS2V** program that is now being made available as source code. It is intended that the scope of the **DSMC** program will be extended in future versions of the book to the extent that it will eventually become an easily applied universal parallel code with capabilities that go beyond those of the widely used **DS2V/3V** codes that have, for some years, been freely downloadable as executables from www.gab.com.au. The extension of **DSMC** to two and three dimensional flows, together with programs for data file production and graphical post-processing will be added in Versions 2 and 3. It is intended that Version 4 will address parallel processing.

Graeme Bird, Sydney, 2013.

gab@gab.com.au

4

Acknowledgements

The readers are thanked in advance for their forbearance of and, hopefully, for their actions in reporting the errors that inevitably remain in this first version. Professor Gennaro Zuppardi of the University of Naples is thanked for his careful proof reading of the early chapters. The publication through CreateSpace will permit a timely rectification of reported errors. The second number in the version designation will increase with each revision.

The author is indebted to the many discussions with and communications from the DSMC community. The biennial International Symposia on Rarefied Gas Dynamics have been particularly influential and the second (Berkeley, 1960) led to the introduction of the DSMC method. More recently, the DSMC workshop meetings in the years between RGD symposia have been similarly valuable. The first of these (Milan, 2003) was organized by the late Carlo Cercignani and the subsequent "Santa Fe meetings" have been organized by Michael Gallis and Alej Garcia. Thanks are due to Sandia National Laboratories for their support of these conferences.

The following is a list of some of the people who have facilitated the author's own work with DSMC over more than fifty years. Some contributions have been through administrative or related support and not everyone in the list has worked with the DSMC codes. There are, of course, many others who have made important contributions to the development and promotion of the DSMC method quite independently of the author.

Tom Fink	*University of Sydney*
Paul Thurston	*Air Force Office of Scientific Research*
Gene Broadwell	*TRW Systems Inc.*
Hans Liepmann	*California Institute of Technology*
John Harvey	*Imperial College, London*
Dan McGregor	*TRW Systems Inc.*
Frank Brock	*NASA Langley Research Center*
Phil Muntz	*University of Southern California*
Jim Moss	*NASA Langley Research Center*
Jim Jones	*NASA Langley Research Center*
Michael Gallis	*Sandia National Laboratories*
Hassan Hassan	*North Carolina State University*
Frank Bergemann	*DLR, Göttingen*
Richard Wilmoth	*NASA Langley Research Center*
Carlo Cercignani	*Politechnic University of Milan*
Doug Auld	*University of Sydney*
Alej Garcia	*San Jose State University*

Special thanks are due to Barbara who indulgently tolerates the author's continuing pursuit of DSMC as an unreasonably time-consuming hobby.

CONTENTS

6 CONTENTS

FUTURE VERSIONS

TWO-DIMENSIONAL FLOW EXTENSIONS

The **DSMC** program code is being extended to deal with two-dimensional and axially-symmetric flows. It will be benchmarked against the existing results from the **DS2V** program in Chapter 8. These and additional studies, together with the source code and associated programs for the generation of the data and the interactive presentation of results, will be included in Version 2. There will be some extensions to program capabilities including options for the inclusion of gravitational fields.

THREE-DIMENSIONAL FLOW APPLICATIONS

The capabilities of the **DSMC** program code will be further extended to deal with three-dimensional flows in a similar manner to the now-obsolete **DS3V** program. Because of the shortcomings in commercial post-processing applications with regard to unstructured data, it will be necessary to also produce auxiliary programs for the interactive display of the results. The three-dimensional extensions are planned be implemented in Version 3, but may be included in Version 2.

PARALLEL PROCESSING

With non-parallel codes on contemporary computers, it is only just possible to extend the Knudsen number range of typical three-dimensional studies into the continuum regime. Once the three-dimensional extensions have been completed, further work will concentrate on taking the maximum advantage of multi-core processors.

LIST OF SYMBOLS

The values of the fundamental physical constants are based on the 2010 CODATA recommended values.

a speed of sound; acceleration; a constant
A area (cross-sectional or planar)
b a constant; miss-distance impact parameter
c_p specific heat at constant pressure
$\mathbf{c}$ molecular velocity
$\mathbf{c_o}$ macroscopic velocity
$\mathbf{c'}$ peculiar or thermal velocity of a molecule
$\mathbf{c}_m$ centre of mass velocity
$\mathbf{c}_r$ relative velocity
C a constant
$\mathbf{C}_p$ diffusion velocity
d molecular diameter
D characteristic linear dimension
D_{12} diffusion coefficient for molecular species 1 and 2
$\mathbf{e}$ unit vector
E molecular energy
f distribution function
f_0 equilibrium distribution function
F cumulative distribution function
$\mathbf{F}$ an external force
F_N real molecules represented by a single DSMC molecule
g_i degeneracy of level i
h Planck constant, 6.6260755×10^{-34} Js; height
H heat of formation of a molecule
k Boltzmann constant, 1.3806488×10^{-23} J/K
k_T thermal diffusion coefficient
K heat conduction coefficient
K_{eq} equilibrium constant
K_n $= \lambda / D$, Knudsen number
l linear dimension; direction cosine with x axis
m molecular mass; direction cosine with y axis
m_r the reduced mass
M $= U/a$, Mach number; a moment

M_S	shock Mach number
n	number density; direction cosine with z axis
n_0	Loschmidt number, 2.6867805×10^{25} /m^3
N	number
$\dot{N}$	number flux
N_A	Avogadro number, $6.02214129 \times 10^{23}$ /mol
N_M	number of real molecules represented by a simulated molecule
p	normal momentum flux or pressure
P	a probability
P_r	$= \mu c_p / K$, Prandtl number
q	heat flux
Q	physical quantity
r	radius
R	$= k / m$, ordinary gas "constant"
R_e	Reynolds number
R_F	random fraction between 0 and 1
R_G	$= k \times N_A$, universal or molar gas constant, 8.3144621 J/mol/K
$\boldsymbol{s}$	position vector
s	$= \beta U$, speed ratio; a distance
S	surface area
S_c	$\mu/(\rho D_{11})$, Schmidt number
t	time
T	temperature
u, v, w	velocity components in the x, y, z directions
U	speed
V	volume
V_m	molar volume, 0.022710953 /m^3
W	width; weighting factor
x, y, z	linear coordinates
z	partition function
Z	relaxation collision number
α	accommodation coeff.; VSS parameter; degree of dissociation
β	inverse of the most probable molecular speed
γ	specific heat ratio
δ	mean space between molecules
ε	molecular energy
Δ	increment
ζ	number of degrees of freedom
η	power law

θ	angle, elevation angle
Θ	characteristic temperature
λ	mean free path
μ	viscosity coefficient
ν	collision rate
ρ	density
σ	total collision cross-section
τ	parallel momentum flux or shear stress
ϕ	azimuth angle, azimuthal impact parameter
χ	deflection angle
ω	viscosity-temperature index; angular velocity
Φ	perturbation term

Superscripts and subscripts

*	post-collision values
coll	associated with a collision
diss	dissociation
el	electronic; surface element
i	inward or incident; quantum level; ionization
int	internal
p,q,s	particular molecular species
r	reflected
ref	reference value
rot	rotational
tr	translational
vib	vibrational
0	equilibrium
1,2	particular values or species

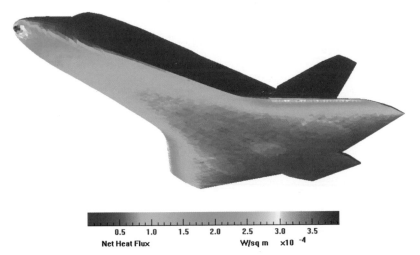

0.5 1.0 1.5 2.0 2.5 3.0 3.5

Net Heat Flux   W/sq m   x10 $^{-4}$

Heat transfer to a re-entering Space Shuttle at an altitude of about 105 km where chemical reactions become significant.
(This was measured on some early flights and that in the nose region of STS-3 at this altitude was in the range 30 to 50 kW/m^2.)

A screen image from a DSMC calculation with the visual interactive **DS3V** program that includes the TCE chemistry model.

Cover:

Mach number contours in a two-dimensional under-expanded jet of a monatomic gas. The pressure at the sonic entry is five times the ambient pressure, while the entry temperature is equal to the ambient temperature. (See §8.13 for details of the DSMC calculation.)

1

OVERVIEW OF THE DSMC METHOD

1.1 Description of the method

The **Direct Simulation Monte Carlo**, or DSMC, method provides a probabilistic physical simulation of a gas flow by simultaneously following the motion of representative model molecules in physical space. The number F_N of real molecules that are represented by each simulated molecule is a computational variable and, while it is a very large number in a rarefied gas, there may be a one-to-one correspondence between simulated and real molecules in a dense gas. A typical computation now involves millions of simulated molecules and requires the computation of billions of intermolecular collisions and surface interactions. Physical time is always an independent variable and any steady flow develops as the large-time state of a physically realistic unsteady flow. Advantage may be taken of flow symmetries in order to reduce the number of independent spatial variables, but collisions and surface interactions are always regarded as three-dimensional events.

The relationships between DSMC and alternative methods at the molecular level were discussed in some detail in Bird (1994). The **Molecular Dynamics**, or MD, method (Alder and Wainwright, 1957) preceded the DSMC method, but it deals with collisions and surface interactions in a deterministic rather than a probabilistic manner. The two approaches are complementary in that the MD method is largely restricted to **dense gases** while the DSMC method requires a **dilute gas** in which the molecular mean free path is much larger than the molecular diameter. The dilute gas assumption together with the probabilistic approach permits the molecular motion and collisions to be uncoupled over a time interval that is small in comparison with the mean collision time. This allows a DSMC calculation to employ the real molecular diameters, but leads to the requirement for a mesh or cell network in physical space. The MD method requires a one-to-one correspondence between real and simulated molecules and, because the number of molecules in a cubic mean free path is inversely proportional to the square of the gas density, its application to a rarefied gas requires the replacement of the large number of realistically small molecules by a smaller number of correspondingly larger molecules.

The analysis of gas flow problems through physical simulation was a radical departure from the traditional approaches that involve the solution of a mathematical model. The strength of the mathematical tradition was such that many people found it hard to accept results that were not obtained from the solution of a mathematical model. The **Boltzmann equation** is the standard mathematical model at the molecular level and, in the earlier years of DSMC development, the question was always "*...but is it a solution of the Boltzmann equation?*" This question was eventually answered by Wagner (1992) who showed that, for time step and grid size tending to zero and molecule number tending to infinity, the simplest form of the DSMC method provides a solution of the Boltzmann equation. Most modern DSMC programs employ more sophisticated procedures and there has consequently been renewed concern about their validity. However, it is important to recognize that the derivation of a mathematical model such as the Boltzmann equation depends on the same physical arguments that have led to the DSMC procedures. A sound physical simulation model should therefore have the same standing as a mathematical model. While analytical comparisons are valuable and instructive, it should not be necessary to demonstrate that the results from the physical simulation model are equivalent to a solution of any mathematical model. This is particularly important with regard to the Boltzmann equation because this equation has restrictions that are not present in DSMC and the method can deal with physical effects that are beyond the Boltzmann formulation.

Physical effects such as chemical reactions and thermal radiation are more readily and more realistically incorporated into a physical simulation model than into a mathematical model. For example, the formulation of the Boltzmann equation assumes the existence of inverse collisions and breaks down if chemical reactions are present. A chemical reaction or the emission of radiation in a DSMC simulation can be event driven and occur either when specified conditions are met in a collision or surface interaction or, in the case of the spontaneous emission of radiation, with a specified probability over any time interval. The DSMC code then becomes more complex and more data is required, but the fundamental level of difficulty is unchanged. When appropriate, the molecular properties in DSMC simulations are expressed as quantum states and this allows more realistic treatments of radiation and chemical reactions.

Continuum methods that assume a continuous medium and continuously distributed variables cannot directly simulate physical effects that occur at the molecular level. The **Navier-Stokes equations** are the standard mathematical model at the continuum level and can deal with chemical reactions or the emission of radiation only through averaged quantities such as rate equations. Moreover, should it be necessary to include the absorption of radiation, there is a drastic increase in the level of mathematical difficulty.

The major disadvantage of DSMC is that the number F_N of real molecules that are represented by each simulated molecule is generally a very large number. This means that the fluctuations in the macroscopic gas properties that are sampled in the simulation are generally much larger than the **fluctuations** in a real gas. If there is an eventual steady state of the flow, the sample size may be built up by taking a **time average** of the properties. In the case of a permanently unsteady flow, many similar runs may be made and the sample may be built up through an **ensemble average** over the runs. The runs must be statistical independent and this is readily attained by employing a different starting random number in each run. The level of the fluctuations then decreases with the square root of the sample size. The fluctuations can generally be reduced to an acceptable level relative to the magnitude of the macroscopic velocities but this becomes difficult when the macroscopic flow speeds are very small. This is because the velocity fluctuations are relative to the molecular velocities that are of the order of the speed of sound in the gas

While the enhanced fluctuations are unphysical when F_N is large, they are physically realistic when the gas is sufficiently dense to allow a one-to-one correspondence between real and simulated molecules. This is another instance of DSMC going beyond the Boltzmann equation because fluctuations are neglected in the Boltzmann model.

Numerical instabilities are absent from applications of the simple form of the DSMC method in which potential collision partners are chosen from anywhere within the computational cell. As with any numerical method that involves a computational grid, the results approach an asymptotic limit as the grid becomes finer. This process is generally described as "convergence", although this word is misleading in that DSMC is not an iterative method. The ratio of the mean separation of the collision partners to the mean free path, or **mcs/mfp ratio** provides the best indication of the degree to which the results will be converged. This ratio is reduced if nearest neighbours are selected as potential collision partners, but the solution is then more sensitive to time step and numerical artifacts have very occasionally been observed when the nearest neighbour option has been selected.

A reduction in the mcs/mfp ratio will always be achieved by the use of an increased number of simulated molecules in order to bring F_N as close as possible to unity. Other than the degree of convergence of the results, the only apparent disadvantage of F_N values above unity in a neutral gas is the enhanced magnitude of the fluctuations. However, in an ionized gas the enhanced fluctuations in the electric field are generally of the same order as the physically significant electric fields. This means that DSMC applications that involve a plasma generally have to be supplemented by an independent computation of the electric field.

There has always been an awareness of the enhanced fluctuations, but the possible presence of stochastic **random walk** effects has often been ignored. It is important that DSMC procedures lead to exact rather than "on the average" conservation of mass momentum and energy. Repeated departures from exact conservation lead to random walks that are more detrimental than the enhanced fluctuations. While the magnitude of the statistical fluctuations declines as the square root of the sample size, both the mean magnitude of the random walk deviations and the mean time between zero crossings increase as the square root of the number of steps. Random walks are most commonly introduced through the radial weighting factors that are generally employed when advantage is taken of cylindrical or spherical symmetry. They also impose limits on the accuracy to which results can be obtained when single precision (32 bit) arithmetic is employed because of the continual rounding of results.

1.2 Molecular and macroscopic flow descriptions

The **macroscopic gas properties** are employed in the continuum models that assume that the gas is a continuous medium. These are the gas **density** ρ, the **pressure** p and the **temperature** T. This is the "thermodynamic" temperature that is assumed to be a scalar quantity with a dimension that is distinct from mass, length and time. The **flow velocity** $\mathbf{c_0}$ is a vector quantity with components u_0, v_0 and w_0 in the x, y and z directions, respectively. The **transport properties** are generally restricted to the coefficients of **viscosity** μ, **heat conduction** K and **forced diffusion** D_{pq}. The subscripts p and q denote the gas species that are associated with the diffusion process. **Thermal diffusion** and **pressure diffusion** are transport phenomena that are almost always neglected in the N-S model. There are additional physical phenomena such as thermal creep that occur only in rarefied gases and are inaccessible to continuum models.

The DSMC method employs the microscopic or **molecular gas properties** that involve a summation over all the molecules* in a volume V. They will be defined for the general case of a gas mixture with s molecular species. The versions for a **simple gas** that is comprised of only one molecular species can be readily written down. When appropriate, the equations will be put into a form that facilitates their evaluation in a DSMC program.

If n_p is the number per unit volume or **number density** of species p, the overall number density n is obtained by summing over all s species. *i.e.*

$$n = \sum_{p=1}^{s} n_p \tag{1.1}$$

The average number $\overline{N}$ of molecules within the volume V is

$$\overline{N} = nV \tag{1.2}$$

and is subject to statistical fluctuations that are described by the Poisson distribution. The probability of a given value of N is

---

* The word "molecules" is used here as a generic term that includes all particles. A distinction will be made between atoms, molecules etc. only when necessary.

$$P(N) = (nV)^N \exp(-nV) / N!. \qquad (1.3)$$

For large values of N, this distribution becomes indistinguishable from the normal or Gaussian distribution

$$P(N) = (2\pi nV)^{-\frac{1}{2}} \exp\left\{(N - nV)^2 / (2nV)\right\} \qquad (1.4)$$

with standard deviation $1/\sqrt{nV}$. The integration of this distribution shows that the probability of an individual number density sample falling within $C\sqrt{nV}$ of the average is $\mathrm{erf}\left(C/\sqrt{2}\right)$, where C is the number of standard deviations. This means that 68.27% of samples will be within one standard deviation, 95.45 within two standard deviations, 99.73 within three standard deviations and 99.994 within four standard deviations. These expectations of the magnitudes of the fluctuations apply to the generally enhanced (by a factor of $\sqrt{F_N}$) fluctuations in a DSMC calculation as well as to those in the real gas.

The number of molecules in one mole of a gas is a physical constant called the **Avogadro number** N_A. Avogadro's law states that the volume occupied by one mole of any ideal gas at **standard temperature** 273.15 K and **standard pressure** 101325 Pa is a constant called the **molar volume** V_m. The number of molecules in one cubic meter at standard temperature and pressure, or **standard number density** n_0 = 2.6867805×10^{23} m^{-3}, is equal to the Avogadro constant divided by the molar volume and is called the **Loschmidt constant**.

The density of a single gas species is the product of the number density and the **mass** m of a single molecule. The macroscopic density is equal to the sum of the individual species densities and can be written

$$\rho = \sum_{p=1}^{s} \left(m_p n_p\right) = n\overline{m}. \qquad (1.5)$$

The macroscopic flow velocity $\mathbf{c_0}$ is related to the momentum flux and is the mass weighted average of the molecular velocities $\mathbf{c}$. i.e.

$$\mathbf{c_0} = \frac{1}{\rho} \sum_{p=1}^{s} \left(m_p n_p \overline{\mathbf{c_p}}\right) = \overline{m\mathbf{c}} / \overline{m}. \qquad (1.6)$$

The **peculiar** or **thermal velocity** c' of a molecule is the difference between the molecular velocity and the mass average or flow velocity. i.e.

$$c' = c - c_0 \tag{1.7}$$

and the mean thermal velocity of a particular species relative to the mass average velocity defines the **diffusion velocity**

$$C_p = \overline{c'_p} = \overline{c_p} - c_0. \tag{1.8}$$

The pressure is related to momentum flux across an element of area in the flow. The momentum flux is in an arbitrary direction and the unit vector normal to the element of area is also in an arbitrary direction. The pressure in the continuum N-S model is assumed to be consistent with Pascal's principle and is a scalar property. On the other hand, because it is a function of two directions, the pressure is a tensor quantity in the molecular model. The definition of the pressure tensor is

$$\mathbf{p} = nm\overline{c'c'} \tag{1.9}$$

and it is best explained in terms of one of the nine components. If u' and v' are the components of c' in the x and y directions, the momentum flux in the x direction across an area element with unit normal vector in the y direction is

$$p_{xy} = nm\overline{u'v'}. \tag{1.10}$$

A **scalar pressure** may be defined as the mean value of the three normal components of the stress tensor. i.e.

$$p = \tfrac{1}{3}nm\overline{(u'^2 + v'^2 + w'^2)} = \tfrac{1}{3}nm\overline{c'^2}. \tag{1.11}$$

Temperatures in the molecular model are measures of the translational and internal energies associated with the molecules so that the **Boltzmann constant** k is effectively non-dimensional. It can be argued that temperature is always a measure of energy, but the thermodynamic notion that "hotness and coldness" is a distinct property of matter still holds sway. To the extent that a distinction is necessary, the temperatures based on molecular energies are sometimes called **kinetic temperatures**.

The translational component of temperature is a measure of the kinetic energy associated with the peculiar or thermal velocities. The equipartition principle states that $\frac{1}{2}kT$ of energy is associated with each degree of freedom and, because there are three degrees of freedom associated with the translational motion of the molecules,

$$\tfrac{3}{2}kT_{tr} = \tfrac{1}{2}\overline{mc'^2} ,$$

or

$$T_{tr} = \tfrac{1}{3}\overline{mc'^2}/k . \qquad (1.12)$$

Then, using Eqn. (1.11),

$$p = nkT_{tr} . \qquad (1.13)$$

The **universal gas constant** R_G is the product of the Boltzmann constant and the Avogadro constant. The ordinary **gas constant** R is equal to the universal constant divided by the molecular weight which, in turn, is the product of the molecular mass and the Avogadro constant. More usefully, the gas constant is equal to the Boltzmann constant divided by the molecular mass so that, for a simple gas, Eqn. (1.13) is equivalent to the perfect gas equation of state.

A DSMC simulation stores the absolute molecular velocity components rather than the thermal velocity components for each simulated molecule and it is desirable to rewrite the temperature definition in terms of the absolute components. The combination of Eqns. (1.7) and (1.12) leads to

$$\tfrac{3}{2}kT_{tr} = \tfrac{1}{2}\overline{mc^2} - \tfrac{1}{2}\overline{mc_0^2} , \qquad (1.14)$$

so that the energy associated with the translational temperature is equal to the kinetic energy of the molecules less the kinetic energy associated with the stream velocity. The use of thermal velocities can also be avoided in the definition of the stress and pressure tensors. For example,

$$p_{xy} = n\left(\overline{muv} - \overline{mu_0 v_0} \right) . \qquad (1.15)$$

In addition the macroscopic flow properties of density, scalar pressure thermodynamic temperature and stream velocity, the continuum equations for compressible flow include the ratio of the specific heat at constant pressure to the specific heat at constant volume. This **specific heat ratio** γ is a measure of the internal energies of the molecules and is related to the average number of effective degrees of freedom $\bar{\zeta}$ of the molecules by

$$\gamma = (\bar{\zeta} + 2) / \bar{\zeta}. \tag{1.16}$$

The energy of each degree of freedom of an internal mode is $\frac{1}{2}kT$ only if the mode is fully excited. Any incomplete excitation is taken into account when setting $\bar{\zeta}$ so that, while the number of physical degrees of freedom is an integer, there may be a non-integer number of effective degrees of freedom. The degree of excitation is a function of temperature and equipartition is not necessarily achieved when the gas is in translational equilibrium.

With the exception of hydrogen and helium at temperatures of just a few degrees Kelvin, the translational mode can be assumed to be fully excited. Rotation can be neglected for monatomic gases and for diatomic molecules about the internuclear axis. This is because the moment of inertia is so small that the angular momentum of the first rotational level would involve energies far higher than those associated with the temperatures of interest. Diatomic molecules have two degrees of freedom associated with the axes normal to the internuclear axis and, with the exception of linear molecules, polyatomic molecules have three degrees of rotation. With the exception of hydrogen, these can be assumed to be fully excited at temperatures of a few tens of degrees Kelvin. By contrast, the vibrational modes of diatomic and polyatomic molecules become fully excited only at some thousands of degrees K.

Separate kinetic temperatures can be assigned to each component of the thermal velocity and to each internal energy mode of each molecular species. The internal energy of a molecule is the sum of the rotational, vibrational and electronic energies. *i.e.*

$$\varepsilon_{int} = \varepsilon_{rot} + \varepsilon_{vib} + \varepsilon_{el} \tag{1.17}$$

The temperature associated with the internal energy is

$$\tfrac{1}{2}\bar{\zeta}kT_{int} = \overline{\varepsilon_{int}} \tag{1.18}$$

and similar expressions could be written for the temperatures associated with the components of the internal energy. The overall temperature is

$$T = (3T_{tr} + \overline{\zeta} T_{int}) / (3 + \overline{\zeta}).$$ (1.19)

The final flow property to be considered is the **heat flux vector** q that is defined by

$$q = \tfrac{1}{2} n m \overline{c'^2 c} + n \overline{\varepsilon_{int} c}.$$ (1.20)

ε_{int} is the internal energy of a single molecule. The component of the heat flux vector on the x direction is

$$q_x = \tfrac{1}{2} n m \overline{c'^2 u'} + n \overline{\varepsilon_{int} u'}.$$ (1.21)

The use of the thermal velocity components can again be avoided by writing this result as

$$q_x = n\left(\tfrac{1}{2} \overline{m c^2 u} - \tfrac{1}{2} \overline{m c^2} u_0 + \overline{\varepsilon_{int} u} - \overline{\varepsilon_{int}} u_0 \right) - p_{xx} u_0 - p_{xy} v_0 - p_{xz} w_0.$$ (1.22)

All temperatures are identical in an equilibrium gas and any differences serve as a quantitative measure of the extent to which a gas departs from equilibrium. The macroscopic or continuum flow properties are averages over the microscopic or molecular properties and more fundamental information on the extent of the departure from equilibrium is provided by the way in which the molecular properties are distributed. A distribution function f is defined for any molecular quantity Q such that the fraction of molecules in the range Q to $Q + dQ$ is

$$dN / N = f dQ.$$ (1.23)

The best known distribution function is the **Maxwellian or equilibrium distribution** f_0 for the thermal speeds of the molecules in an equilibrium gas at temperature T. This is generally written

$$f_0 = (\beta^3 / \pi^{3/2}) \exp(-\beta^2 c'^2),$$ (1.24)

where β is the reciprocal of the most probable molecular speed that is related to the temperature by

$$\beta = (2kT / m)^{-1/2}. \tag{1.25}$$

The **speed ratio**

$$s = U\beta \tag{1.26}$$

is generally preferred to the **Mach number** as the non-dimensional speed parameter. It is smaller than the Mach number by $(2 / \gamma)^{1/2}$.

The macroscopic properties associated with a fluid element change as the element traverses a fluid flow. The distributions of the translational velocity and the energy in each internal mode also change. The idealized case in which the changes are sufficiently slow for all the distributions to effectively attain the equilibrium form at all times and at all locations corresponds to an **isentropic** flow in the continuum description. The **entropy** involves $f \ln f$ and is difficult to sample from the microscopic properties, but the distribution functions provide a more precise description of the degree of non-equilibrium.

1.3 The need for a molecular description

As outlined in the preceding section, the Navier-Stokes, or N-S, equations ignore the discrete molecular nature of a real gas and assume that the gas is a continuous medium. The conservation of mass, momentum and energy is enforced in both continuum and particle methods, but the N-S transport terms that include the coefficients of viscosity, heat conduction and diffusion break down for sufficiently large values of the **Knudsen number** K_n. This number is defined as the ratio of the molecular mean free path λ to a characteristic linear dimension D. That is,

$$K_n = \lambda / D. \tag{1.27}$$

An **overall Knudsen number** is obtained if the typical dimension is set to an overall flow feature such as the diameter or length of a tube or the size of a body. Overall Knudsen numbers have traditionally been employed to define flow regimes such as **continuum, slip, transition and free-molecule**. A continuum flow is one in which the transport terms in the Navier-Stokes equations are everywhere valid. The free-molecule or collisionless regime is the opposite extreme in which there are no intermolecular collisions. Slip flow is a hypothetical regime in which it is assumed that the continuum equations remain valid as long as allowance is made for velocity and temperature slips at solid boundaries. The transition regime is everything between slip flow and free-molecule flow. Because of the arbitrary choice of the typical dimension, these classifications have often proved to be misleading and a single flow may contain regions that fall into two or more different regimes. In particular, the traditional criteria served to excessively minimize the transition regime. It is preferable to employ **local Knudsen numbers** based on the scale lengths of the gradients in the flow properties.

Local Knudsen numbers provide definitive information on whether or not the N-S model is valid. This is because the transport terms are valid only if the distribution of molecular velocities f is a small perturbation of the equilibrium or Maxwellian distribution f_0. The Navier-Stokes terms that involve the viscosity, heat conduction and diffusion coefficients were essentially empirical until they were linked to the microscopic properties through the Chapman-Enskog theory (Chapman and Cowling, 1970). For the special case in which there is a velocity gradient only in the y direction and a temperature gradient only in the x direction, the Chapman-Enskog distribution can be written

$$f = f_0\left[1 + C\beta v'\left\{3\left(\beta^2 c'^2 - \frac{5}{2}\right)\frac{\lambda}{T}\frac{\partial T}{\partial x} + 4\beta u's\frac{\lambda}{u_0}\frac{\partial u_0}{\partial y}\right\}\right], \qquad (1.28)$$

where C is a constant of order unity that depends on the composition of the gas. The local Knudsen numbers based on the scale lengths of the relevant macroscopic gradients appear explicitly in the perturbation term. All the coefficients of these local Knudsen numbers, other than the speed ratio that appears in the term related to the velocity gradient, can be regarded as being of order unity.

The viscosity coefficient can be related to the mean free path and, for a gas with viscosity coefficient proportional to temperature to the power ω, the Knudsen, Mach and Reynolds numbers are related by

$$K_n = (2\gamma/\pi)^{1/2}[(5 - 2\omega)(7 - 2\omega)/15]M_a/R_e. \qquad (1.29)$$

The numerical value of the term in black type ranges from about 1.5 to 2.

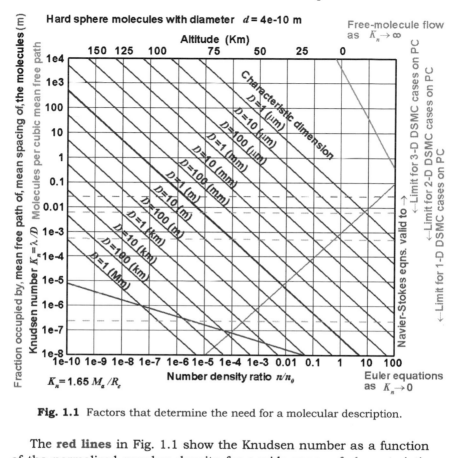

Fig. 1.1 Factors that determine the need for a molecular description.

The **red lines** in Fig. 1.1 show the Knudsen number as a function of the normalized number density for a wide range of characteristic linear dimensions. The density is also expressed in terms of altitude. The calculations on which the figure is based employed a hard sphere molecular model with a diameter of 0.4 nm and any significant departure from these guidelines would most likely be related to differences in the effective collision cross-section.

The mean free path in a hard sphere gas is $1/(\sqrt{2}\pi d^2 n)$, so that the number N of molecules in a cubic mean free path is

$$N \equiv n\lambda^3 = \left(2^{3/2}\pi^2 d^6 n_0^2\right)^{-1}\left(n_0/n\right)^2. \qquad (1.30)$$

The coefficient of the density ratio term is 3860 for the molecular diameter in Figure 1 and N is plotted as the **green line**. The minimum size of flow features is of the order of the mean free path, so that the number of molecules that are associated with a particular feature decreases as the square of the number density. This has often been found to be a counter-intuitive result.

The fraction of the volume that is occupied by the molecules is the product of the number density and the volume of a molecule and this is shown by the **blue line** in Figure 1. The mean spacing between the molecules δ is equal to the inverse cube root of the number density and is shown in meters as the **mauve line**. The mean free path in metres is equal to the Knudsen number when the characteristic dimension is one metre and this line is duplicated as a **purple line**. A slight extrapolation of the mean spacing line shows that, at standard number density, the mean free path is about an order of magnitude greater than the mean spacing between the molecules. This, in turn, is an order of magnitude larger than the molecular diameter so that the gas at standard density just qualifies as a dilute gas. At densities greater than two or three times the standard density, the gas has to be regarded as a dense gas. The volume occupied by the molecules must then be taken into account, the perfect gas equation of state no longer applies, even in a monatomic gas, and it can no longer be assumed that all collisions are binary collisions.

The errors due to the assumption of a continuous fluid become significant when the Knudsen number (either overall or local, as appropriate) is above 0.02 or 0.03. Continuum models such as the Navier-Stokes equations then cease to provide a reasonably accurate representation of the flow and they must be replaced by a molecular model.

1.4 Computational considerations

The computational requirements for DSMC calculations are highly problem dependent, but Figure 1.1 includes a rough guide to the smallest Knudsen number that is attainable in single-processor calculations on contemporary personal computers. There are separate limits for one, two and three-dimensional flows and, in all cases, there is an overlap with the upper Knudsen number limit for the validity of the Navier-Stokes equations. The estimates are most likely optimistic in that they do not fully account for the fact that, especially for unsteady flows, the duration of the physical flow time that must be simulated increases as the Knudsen number decreases. For example, while there is no limit on the number density in homogeneous gas calculations, the time-scales associated with chemical reactions may impose practical limits on DSMC simulations.

It has been possible to set limits for single processor calculations only because the maximum CPU "clock speed" has leveled off at about 3 GHz. This means that parallel computation is now required in order to take advantage of the latest computer hardware. Most of the existing parallel DSMC codes employ domain decomposition in which the flowfield is divided into multiple regions such that each processor deals with a single region. The overhead associated with moving the simulated molecules between regions means that the computing time does not decrease linearly with the number of processors and the gains level off at the order of 30 to 100 regions. The number of cores in graphics processing units, or GPUs, is generally far more than the optimum number for domain decomposition. There have been several DSMC implementations (*e.g*, Gladkov *et. al.*, 2012 and Su *et. al.*, 2012) that employ the Nvidia GPU with the CUDA programming system and speed gains of two orders of magnitude have been claimed. The rate of increase in the number of processing cores in mainstream computer CPU's has been disappointingly slow, but can be expected to soon catch up with the number of processors in the clusters that have been employed for most applications of the programs that implement domain decomposition. The optimum use of this capability will almost be through splitting the program loops between processors rather than through domain decomposition that effectively splits the whole program between processors. This will be facilitated by the new computer language constructs such as the co-arrays that have been introduced into the recent Fortran specifications. Co-arrays and the similar constructs in the C languages are now being implemented in the compilers and will hopefully allow future DSMC programs to more easily take full advantage of multi-core processors. While the processor speeds have leveled off, the memory capacity of computers has continued to increase. While early DSMC applications were restricted by both memory size and the practical limit on the time required for the calculation, a typical workstation can now cope with a billion simulated molecules and the computation time is now the dominant issue.

The discussion in the preceding paragraph assumes that the objective of parallel processing is to extend the Knudsen number limit for a problem that would require a run of prohibitive length. If the application has a Knudsen number that is within the limits for single processor calculations and the need is to just increase the sample size, there is a much simpler way to take full advantage of multiple processors. Multiple images of the program can be run with different random number sequences and an ensemble average made of the results. The averaging may be at the macroscopic or microscopic levels. If it is the macroscopic variables that are averaged over the runs, the signal-to-noise ratio is improved but the fluctuations are unphysical. On the other hand, the averaging of the molecular or microscopic properties is equivalent to making the calculation with a larger number of molecules. It may then be possible to achieve a one-to-one correspondence between real and simulated molecules so that the fluctuations match those in the real gas.

The green line in Figure 1 shows that only five thousand molecules per cubic mean free path are required at standard temperature and pressure in order to make a calculation with physically realistic fluctuations. It is often claimed that, because the flow speeds in micro-devices are generally very small in comparison with the molecular speeds, DSMC must be replaced by alternative approaches. However, with ensemble averaging at the molecular level, it is feasible to make DSMC calculations with $F_N = 1$ and therefore with physically realistic fluctuations. Because the alternative approaches invariably neglect the fluctuations, DSMC calculations are to be preferred.

In any DSMC simulation of a three-dimensional flow, there is a total number of simulated molecules for which the ratio F_N is unity. For a flowfield of volume V with an average number density $\bar{n}$, the number of simulated molecules for the fluctuations to be realistic is

$$N_m = \bar{n}V. \tag{1.31}$$

If the number of simulated molecules has to be set to less than this value, the fluctuations are unphysically large. Conversely, the number should not be allowed to exceed this value because the fluctuations would then be unphysically small.

On the other hand, the cross-section of a one-dimensional flow and the width of a two-dimensional flow simulation are subject to arbitrary choice and have generally been set to unity within DSMC programs.

Similarly, the azimuthal width of an axially-symmetric flow is usually set to 2π. However, for one and two-dimensional flows, it is possible to write down the flow cross-section or width for which $F_N = 1$ and at which the fluctuations become physically realistic. In a one-dimensional flow of length l with a total N_M simulated molecules and an average number density $\bar{n}$, the cross-sectional area A at which the fluctuations are realistic averages over the area is

$$A = N_M / (\bar{n}\, l). \qquad (1.32)$$

Similarly, for a two-dimensional flowfield of area A_f, the flow width at which the fluctuations are realistic averages over the width is

$$W = N_M / (\bar{n}\, A_f). \qquad (1.33)$$

An axially-symmetric flowfield of area A_f can be regarded as a three-dimensional flow with volume $V = 2\pi r_c A_f$, where r_c is the radius of the centroid of A_f. Equation (1.31) can then be used to determine the total number of simulated molecules for physically realistic fluctuations. With fewer than this number of simulated molecules, the fluctuations are physically realistic averages across a slice of azimuthal extent

$$\Delta\phi = N_M / (\bar{n}\, r_c\, A_f). \qquad (1.34)$$

The symmetry path varies with radius and the sample of molecules within an element of area is proportional to the radius of the element. The small sample near the axis is a disadvantage in some DSMC applications and a weighting factor proportional to the radius is frequently applied to nullify the effect of radius on the sample size. This leads to fractional molecule duplication if a simulated molecule moves to a smaller radius and a fractional removal if it moves to a larger radius. However, as noted earlier, this gives rise to random walks and radial weighting factors should be avoided whenever possible.

The DSMC procedures that were employed up to the early 90's were described in Bird (1994) and were illustrated by a set of demonstration codes. The two and three-dimensional demonstration codes employed simple rectangular cell grids and were intended only for applications that involve surfaces that lie on cell boundaries. These have occasionally been modified for application to cases with surfaces that cut across cell boundaries. The results are poor in comparison with those from programs that employ adaptive body-fitted cells, but it is undesirable to burden DSMC with the overhead that is associated with grid generation in continuum CFD. This can be avoided through the procedures that have been employed in the DS2V/3D programs (Bird, 2007). A feature of these programs is that the only computational variable that is required in the data specification is the number of megabytes that are required at the start of the calculation. The cell structure is generated automatically and can be continuously adapted to a uniform number of simulated molecules per cell. There are separate collision and sampling cells and every molecule, as well as every collision cell, is assigned a time parameter. These parameters are kept up to date with the physical flow time through time steps that vary over the flowfield and are continuously adapted to the local mean collision time. Nearest-neighbour collisions can be employed and additional molecular models have been introduced. The procedures that are described and implemented in the later chapters further extend these newer DSMC developments.

1.5 DSMC program structure

The flowchart of typical DSMC programs has not changed greatly since the method was first introduced, but the modern features have necessitated some modifications. A flowchart that applies to both the **DSMC** and **DS2V** programs is shown in Fig. 1.2. The molecule moves and collisions are uncoupled over a small time step that is set to a fixed fraction of the local mean collision time. The density in some applications varies by several orders of magnitude over the flowfield. The region of highest density is often the most important and a good calculation requires that the time step be small in comparison with the mean collision time in that region. An older program with a fixed time step is then very inefficient and a variable time step is now employed. However, there is still some penalty because the main loop must be over the minimum time step. This appears in the green process block in the flow chart, while the local time steps that may be very much larger appear in the orange process blocks. Note that the molecules move and collisions are calculated in a collision cell when their times fall more than half a time step behind the overall time step. This means that the average molecule and collision times are, at all times, equal to the flow time.

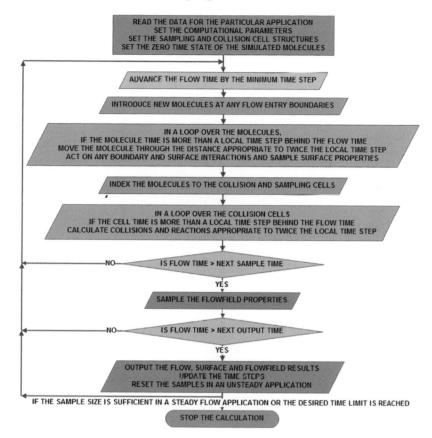

Fig. 1.2 The flow chart for the **DSMC** and **DS2V** programs .

The time step was a data item in the early programs, but is now set automatically by the programs. It is initially set to the value appropriate to the stream or reference gas and is updated to the current flow conditions whenever results are output. The sampling intervals are set as multiples of the time step and the output interval to multiples of the sampling intervals. The multiples are set to default values that may be altered through the adjustable computational parameters in the source code. The only computational parameter in the data files is the initial number of megabytes to be used in the calculation, but the computational parameters within the source code allow users to make adjustments that hopefully lead to optimal settings for special cases.

References

Alder, B. J. and Wainwright, T. E. (1957). Studies in Molecular Dynamics, *J Chem. Phys.* **27**, 1208-1209.

Bird, G. A. (1994) *Molecular Gas Dynamics and the Direct Simulation of Gas Flows,* Clarendon Press, Oxford.

Bird, G. A. (2007). *Sophisticated DSMC,* Short course presented at the DSMC07 meeting, Santa Fe. (downloadable from www.gab.com.au).

Chapman, S. and Cowling, T.G. (1970). *The Mathematical Theory of Non-uniform Gases* 3rd edn., Cambridge University Press.

Gladkov, V., Tapia, J-J, Alberts, S. and D'Souza, R. (2012). Graphics processing unit based direct simulation Monte Carlo, *SIMULATION* **88**, 680-693.

Su, C.C., Smith, M. R., Wu, J.S., Hseih, C.W., Tseng, K.C. and Kuo, F. A. (2012). Large-Scale Simulations on Multiple Graphics Processing Units (GPUs) for the Direct Simulation Monte Carlo method, J. Comput. Phys, **231**, 7932=7958.

Wagner, W. (1992). A convergence proof for Bird's direct simulation Monte Carlo method for the Boltzmann equation, *J. Stat. Phys.* **66**,1011–1044.

2

REFERENCE STATES AND BOUNDARY CONDITIONS

2.1 Spatial quantities in an equilibrium gas

As noted in §1.3, the distribution function for the thermal speeds of the molecules in an equilibrium or Maxwellian gas at temperature T is

$$f_0 = \left(\beta^3/\pi^{3/2}\right)\exp\left(-\beta^2 c'^2\right), \tag{2.1}$$

where

$$\beta = (2RT)^{-\frac{1}{2}} = (2kT/m)^{-\frac{1}{2}}. \tag{2.2}$$

This means that the fraction of molecules within a velocity space element $d\mathbf{c}' = du'\,dv'\,dw'$ located at $\mathbf{c}'$ is

$$\frac{dN}{N} - \frac{dn}{n} = \left(\beta^3/\pi^{3/2}\right)\exp\left(-\beta^2 c'^2\right)du'\,dv'\,dw'. \tag{2.3}$$

The peculiar or thermal velocity $\mathbf{c}'$ is equal to $\mathbf{c} - \mathbf{c}_0$ so that, in **Cartesian coordinates** (u,v,w), the fraction of molecules with velocity components from u to $u+du$, v to $v+dv$ and w to $w+dw$ is

$$\frac{dn}{n} = \left(\beta^3/\pi^{3/2}\right)\exp\left[-\beta^2\left\{(u-u_0)^2 + (v-v_0)^2 + (w-w_0)^2\right\}\right]du\,dv\,dw. \tag{2.4}$$

For **polar coordinates** (c',θ,ϕ) in a frame of reference moving with the stream velocity, the volume of the velocity space element is

$$c'^2 \sin\theta\,d\theta\,d\phi\,dc'$$

The fraction of molecules with speed between c' and dc', which have an elevation angle between θ and $d\theta$, and an azimuth angle between ϕ and $d\phi$ is, therefore,

$$\frac{dn}{n} = \left(\beta^3 / \pi^{3/2}\right)c'^2\exp\left(-\beta^2 c'^2\right)\sin\theta\, d\theta\, d\phi\, dc'. \qquad (2.5)$$

The fraction of molecules with speeds between c' and $c' + dc'$, irrespective of direction, is obtained from Eqn. (2.5) by integrating ϕ between the limits 0 to 2π and θ from 0 to π, to give

$$\frac{dn}{n} = \left(4 / \pi^{1/2}\right)\beta^3\, c'^2\exp\left(-\beta^2 c'^2\right)dc'\, .$$

A distribution function $f_{c'}$ may be defined such that the fraction of molecules with speeds between c' and $c' + dc'$ is $f_{c'}\, dc'$. Therefore,

$$f_{c'} = \left(4 / \pi^{1/2}\right)\beta^3\, c'^2\exp\left(-\beta^2 c'^2\right). \qquad (2.6)$$

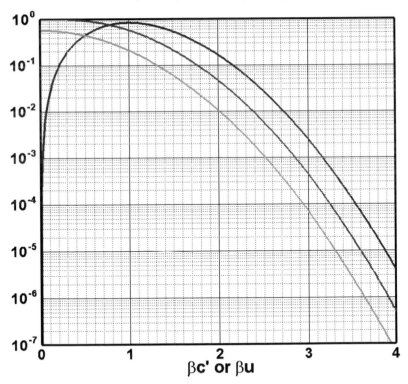

Fig. 2.1 The **equilibrium distribution functions** for the **molecular speed** and for a molecular velocity component in a stationary gas. Also, the **fraction of molecules with speed above c'** in an equilibrium gas.

There is zero possibility that a molecule is stationary and, because β is the inverse of the **most probable molecular speed** $c'_m = \sqrt{2kT/m}$, the distribution function $f_{c'}$ has a maximum when $\beta c'$ is unity. The **average molecular speed** is (Bird, 1994)

$$\overline{c'} = 2\big/\big(\pi^{1/2}\beta\big) = \big(2\big/\pi^{1/2}\big)c'_m \tag{2.7}$$

and the fraction of molecules with thermal speeds above c' is

$$\frac{dn}{n} = 1 + \big(2\big/\pi^{1/2}\big)\beta\, c'\exp\big(-\beta^2 c'^2\big) - \mathrm{erf}\big(\beta\, c'\big). \tag{2.8}$$

The fraction of molecules with a velocity component within a given range, irrespective of the magnitude of the other components, is obtained by integrating Eqn. (2.3) over the other components. This leads to the following expression for the distribution function of a thermal velocity component

$$f_{u'} = \big(\beta\big/\pi^{1/2}\big)\exp\big(-\beta^2 u'^2\big). \tag{2.9}$$

This is plotted in Fig. 2.1 for the molecules with a positive velocity component. The equilibrium distribution function is spherically symmetric about the point representing the stream velocity, the most probable value of a peculiar velocity component is zero. The average of the positive velocity components is , using Eqn. (2.7),

$$\overline{u'} = 1\big/\big(\pi^{1/2}\beta\big) = \overline{c'}\big/2. \tag{2.10}$$

The equilibrium distribution function for the energy ε_{int} in an internal mode with ς degrees of freedom is

$$f_{\varepsilon_{int}} \propto \varepsilon_{int}^{\varsigma/2-1}\exp\big\{-\varepsilon_{int}\big/(kT)\big\}. \tag{2.11}$$

Should the continuously distributed internal energy of the classical models be replaced by **quantum levels** 0 to j, the equilibrium or **Boltzmann distribution** is such that the fraction of molecules in level i with energy ε_i is

$$\frac{N_i}{N} = \frac{g_i\exp\big\{-\varepsilon_i\big/(kT)\big\}}{\sum_{i=0}^{j} g_i\exp\big\{-\varepsilon_i\big/(kT)\big\}}, \tag{2.12}$$

where the **degeneracy** g_i is the number of states at energy level i.

For a gas mixture in equilibrium, the equations of this section may be applied separately to each of the molecular species in the mixture.

2.2 Fluxal quantities in an equilibrium gas

Consider the flux of molecular quantities across an element in an equilibrium gas. The stream velocity c_0 is inclined at angle θ to the unit normal vector e of the element. Without loss of generality, Cartesian coordinates can be set such that the unit normal vector is in the negative x direction, the steam velocity is in the x-y plane, and the element is in the y-z plane. The velocity components of a molecule are then

$$u = u' + c_0 \cos\theta$$

$$v = v' + c_0 \sin\theta \qquad (2.13)$$

and $$w = w'.$$

The inward flux of some quantity Q is in the x direction and is given by

$$n\overline{Qu} \quad \text{or} \quad n\int_{-\infty}^{\infty}\int_{-\infty}^{\infty}\int_{0}^{\infty} Q u f\, du\, dv\, dw.$$

Eqns. (1.26), (2.4), and (2.13) allow this to be written

$$\frac{n\beta^3}{\pi^{3/2}} \int_{-\infty}^{\infty}\int_{-\infty}^{\infty}\int_{-c_0\cos\theta}^{\infty} Q(u' + c_0\cos\theta)\exp\left\{-\beta^2\left(u'^2 + v'^2 + w'^2\right)\right\} du'\, dv'\, dw'. (2.14)$$

The **inward number flux** $\dot{N}_i$ is obtained by setting Q to unity in Eqn. (2.14). The integrals may be evaluated to give

$$\beta\dot{N}_i/n = \left[\exp\left(-s^2\cos^2\theta\right) + \pi^{1/2}s\cos\theta\{1 + \mathrm{erf}(s\cos\theta)\}\right]/(2\pi^{1/2}). \quad (2.15)$$

For a stationary gas, the stream speed and the speed ratio are zero so that, using Eqn. (2.7) and noting that $c' = c$ in a stationary gas,

$$\dot{N}_i = nc_m/(2\pi^{1/2}) = n\overline{c}/4. \qquad (2.16)$$

This is as expected because half the gas is moving towards the element with the average speed $\overline{c}/2$.

The **inward normal momentum flux** or pressure p_i is obtained by setting Q to mu or $m(u' + c_0\cos\theta)$. The final result is

$$\frac{\beta^2 p_i}{\rho} = \frac{s\cos\theta\exp\left(-s^2\cos^2\theta\right) + \pi^{1/2}\left\{1 + \text{erf}\left(s\cos\theta\right)\right\}\left(\frac{1}{2} + s^2\cos^2\theta\right)}{2\pi^{1/2}} \quad (2.17)$$

For a stationary gas, this becomes

$$p_i = \rho/\left(4\beta^2\right) = \rho RT/2 - p/2, \quad (2.18)$$

as would be expected in an equilibrium gas because the inward moving molecules contribute half the pressure.

The **inward parallel momentum flux** or **shear stress** τ_i is obtained by setting Q to mv or $m(v' + c_0 \sin\theta)$. The final result is

$$\beta^2 \tau_i /\rho = s\sin\theta\left(\beta\dot{N}_i / n\right) \quad (2.19)$$

and the shear stress is, of course, zero in a stationary gas.

The **inward translational energy flux** $q_{i,tr}$ to the element is obtained by setting Q to $\frac{1}{2}mc^2$ or $\frac{1}{2}m(u^2 + v^2 + w^2)$. This leads to

$$\frac{\beta^3 q_{i,tr}}{\rho} = \frac{\left(s^2 + 2\right)\exp\left(-s^2\cos^2\theta\right) + \pi^{1/2}s\cos\theta\left\{1 + \text{erf}\left(s\cos\theta\right)\right\}\left(\frac{1}{2} + s^2\right)}{4\pi^{1/2}}$$
$$(2.20)$$

For a stationary gas,

$$\beta^3 q_{i,tr} /\rho = 1/\left(2\pi^{1/2}\right). \quad (2.21)$$

Unlike the number and momentum fluxes, the energy flux is affected by the presence of internal energy. However, the internal energy is independent of the molecular velocity and the **inward internal energy flux** is simply the product of the number flux and the average internal energy. *i.e.*

$$q_{i,int} = \dot{N}_i \overline{\varepsilon_{int}}. \quad (2.22)$$

For gas mixtures, the properties are summed over the species.

Note that **"continuum thinking" does not necessarily apply at the molecular level.** Consider a collisionless gas with a bimodal distribution of similar molecules denoted by subscripts 1 and 2. From Eqn. (1.6), the stream velocity is proportional to $n_1 u_1 + n_2 u_2$, while Eqn. (2.19) shows that the shear stress is proportional to $n_1 u_1 \sqrt{T_1} + n_2 u_2 \sqrt{T_2}$. Therefore, for opposing component velocities, the shear stress is in the opposite direction to the stream velocity if $\sqrt{T_2/T_1} > \left|n_1 u_1/(n_2 u_2)\right|$. Similarly, the heat flux can be from a cold to a hot region and Fourier's law is not always valid at the molecular level.

2.3 Collisional quantities in an equilibrium gas

The binary **collision rate** in a gas is

$$\nu = n\overline{\sigma c_r} . \tag{2.23}$$

Here, σ is the **total collision cross-section** and c_r is the magnitude of the **relative velocity** $\mathbf{c}_r = \mathbf{c}_1 - \mathbf{c}_2$ between the two molecules. The cross-section is a function of the relative velocity and the translational energy in a collision is proportional to the square of c_r. When dealing with collision energies, it is customary to introduce the **reduced mass**

$$m_r = m_1 m_2 / (m_1 + m_2). \tag{2.24}$$

It is therefore desirable to obtain an expression for the mean value of the relative speed raised to the arbitrary power b.

With the assumption of **molecular chaos**, the two-particle distribution function is the product of the two single-particle distribution functions f_1 and f_2. The required **mean value over the molecules** is then

$$\overline{c_r^b} = \int\limits_{-\infty}^{\infty} \int\limits_{-\infty}^{\infty} c_r^b f_1 f_2 \, d\mathbf{c}_1 d\mathbf{c}_2.$$

With both the single-particle distributions set to the equilibrium distribution, this may be transformed to the following integral over c_r

$$\left(4/\pi^{1/2}\right)\left\{m_r/(2kT)\right\}^{3/2} \int\limits_{0}^{\infty} c_r^{b+2} \exp\left\{-m_r c_r^2/(2kT)\right\} dc_r. \tag{2.25}$$

This may be evaluated to give

$$\overline{c_r^b} = \left(2/\pi^{1/2}\right) \Gamma\left\{(b+3)/2\right\}\left(2kT/m_r\right)^{b/2}. \tag{2.26}$$

This result applies to collisions between a molecule of species 1 with a molecule of species 2. For a simple gas comprised of a single molecular species, the reduced mass is $m/2$. The mean value of the relative speed is in a simple gas is then obtained by setting b to unity, i.e.

$$\overline{c_r} = 2^{3/2}/\left(\pi^{1/2}\beta\right) = 2^{1/2}\overline{c'}. \tag{2.27}$$

The **variable hard sphere**, or **VHS**, and the **variable soft sphere**, or **VSS**, molecular models are the dominant molecular models in applications of the DSMC method. These have a cross-section that is proportional to a power of the relative speed in a collision. For reasons that will become clear in the following section, the power will be set to $-(2\omega-1)$. The proportionalities between the quantities can then be written

$$\sigma \propto d^2 \propto c_r^{-(2\omega-1))} \propto E_{tr}^{-(\omega-1/2)} \propto T^{-(\omega-1/2)}. \tag{2.28}$$

The collision rate in a VHS gas is therefore proportional to $c_r^{2(1-\omega)}$ and, using Eqn. (2.26),

$$\overline{c_r^{2(1-\omega)}} = \left(2/\pi^{1/2}\right)\Gamma\left(\tfrac{5}{2}-\omega\right)(2kT/m_r)^{1-\omega}. \tag{2.29}$$

The **mean value over the collisions** in an equilibrium VHS gas of $c_r^{b'}$ may be obtained by setting $b = 2(1-\omega)$ in Eqn. (2.25), including $c_r^{b'}$ within the integral, and dividing by the integral with $b' = 0$. *i.e.*

$$\overline{c_r^{b'}} = \int_0^\infty c_r^{b'+2(2-\omega)}\exp\left(-\frac{m_r c_r^2}{2kT}\right)dc_r \bigg/ \int_0^\infty c_r^{2(2-\omega)}\exp\left(-\frac{m_r c_r^2}{2kT}\right)dc_r$$

or

$$\overline{c_r^{b'}} - (2kT/m_r)^{b'/2}\,\Gamma\left(\tfrac{1}{2}b' + \tfrac{5}{2} - \omega\right)/\Gamma\left(\tfrac{5}{2}-\omega\right). \tag{2.30}$$

The mean value of the translational energy in the collision $E_{tr} = \tfrac{1}{2}m_r c_r^2$ may be evaluated by setting $b' = 2$ in Eqn. (2.30) to give

$$\overline{E_{tr}} = \left(\tfrac{5}{2}-\omega\right)kT. \tag{2.31}$$

A VHS reference diameter d_{ref} may be defined as the average molecular diameter in collisions in a gas at the reference temperature T_{ref}. Equation (2.28) shows that d^2 is proportional to $c_r^{-(2\omega-1)}$ and

$$\overline{c_r^{2\omega-1}} = (2kT/m_r)^{\omega-1/2}\big/\Gamma\left(\tfrac{5}{2}-\omega\right),$$

so that

$$d = d_{ref}\left[\left\{2kT_{ref}\,/\left(m_r c_r^2\right)\right\}^{\omega-1/2}\Big/\Gamma\left(\tfrac{5}{2}-\omega\right)\right]^{1/2}. \tag{2.32}$$

The collision rate of Eqn. (2.23) can therefore be written for an equilibrium VHS gas as

$$v_0 = n \pi d_{ref}^2 \left(2kT_{ref}/m_r\right)^{\omega-1/2} \overline{c_r^{2(1-\omega)}}/\Gamma\left(\tfrac{5}{2}-\omega\right).$$

Then, using Eqn. (2.29),

$$v_0 = 2\sqrt{\pi}\, n\, d_{ref}^2 \left(2kT_{ref}/m_r\right)^{1/2} \left(T/T_{ref}\right)^{1-\omega}. \tag{2.33}$$

The mean free path is defined in the frame of reference moving with the stream velocity and is equal to the mean thermal speed given in Eqn. (2.7) divided by the collision frequency. For a simple gas with $m_r = m/2$

$$\lambda_0 = 1 \Big/ \left\{\sqrt{2}\, \pi d_{ref}^2\, n\left(T_{ref}/T\right)^{\omega-1/2}\right\}. \tag{2.34}$$

Note that the collision rate is independent of temperature in a Maxwell gas $(\omega = 1)$ and the mean free path is independent of temperature in a hard sphere gas $(\omega = \tfrac{1}{2})$. The value of ω in real gases is well between these limits and both the collision rate and the mean free path are functions of temperature. Because the inverse power law models require the definition of arbitrary cut-offs and do not lead to unambiguous values of the collision rate and mean free path, the hard sphere results have habitually been employed in data reduction. This has led to errors in data sets that cover a wide range of temperature and the temperature dependent cross-section of the VHS model could have provided a practical solution to this problem.

The fraction of collisions in which E_{tr} exceeds some reference value E_a is important in the context of chemical reactions. The total number of collisions is proportional to the integral in Eqn. (2.25) with $b = 2(1-\omega)$ and the required fraction is the ratio of the integration from E_a to the integration from 0 that is given by Eqn. (2.26). *i.e.*

$$\frac{dN}{N} = \frac{2}{\Gamma\left(\tfrac{5}{2}-\omega\right)}\left(\frac{m_r}{2kT}\right)^{5/2-\omega} \int\limits_{(2E_a/m_r)^{1/2}}^{\infty} c_r^{2(2-\omega)} \exp\left(-\frac{m_r c_r^2}{2kT}\right) dc_r$$

or, if the incomplete Gamma function is introduced,

$$dN/N = \Gamma\left\{\tfrac{5}{2}-\omega, E_a/(kT)\right\}/\Gamma\left(\tfrac{5}{2}-\omega\right). \tag{2.35}$$

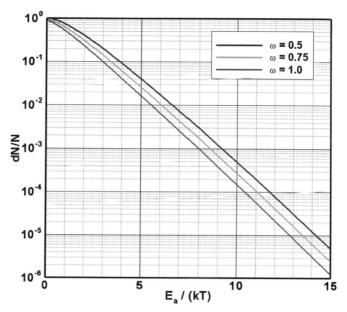

Fig. 2.2 The fraction of collisions in an equilibrium gas that have relative translational energies above a specified reference energy .

While the preceding equations in this section have allowed for molecules of different mass, the results have essentially been for a simple gas. Consider a mixture of s separate molecular species of VHS molecules that are specified by an s by s array of reference diameters, reference temperatures, and ω parameters. The collision rate result of Eqn. (2.32) is readily modified to give the collision rate of molecules of p molecular species with molecules of q species. This is

$$\left(v_{pq}\right)_0 = 2\pi^{1/2} n_q \left(d_{ref}\right)_{pq}^2 \left(2k\left(T_{ref}\right)_{pq}/m_r\right)^{1/2} \left(T/\left(T_{ref}\right)_{pq}\right)^{1-\omega_{pq}} . \quad (2.36)$$

The mean collision rate for species p is obtained by summing over all species

$$\left(v_p\right)_0 = \sum_{q=1}^{s} \left\{ 2\pi^{1/2} n_q \left(d_{ref}\right)_{pq}^2 \left(2k\left(T_{ref}\right)_{pq}/m_r\right)^{1/2} \left(T/\left(T_{ref}\right)_{pq}\right)^{1-\omega_{pq}} \right\} . \quad (2.37)$$

The mean collision rate per molecule for the mixture is obtained as the sum over the p species weighted by the number fraction. *i.e.*

$$\left(v\right)_0 = \sum_{p=1}^{s}\left\{\left(n_p/n\right)\left(v_p\right)_0\right\}$$

(2.38)

The total number of collision events is

$$N_c = \tfrac{1}{2}nv\;.$$

(2.39)

The symmetry factor of one half is included because the summations count each collision event twice.

The mean free path of a species p molecule is

$$\left(\lambda_p\right)_0 = 1\bigg/\sum_{q=1}^{s}\left\{\pi\left(d_{ref}\right)_{pq}^2 n_q\left(\left(T_{ref}\right)_{pq}/T\right)^{\omega_{pq}-1}\left(1+m_p/m_q\right)^{1/2}\right\}$$

(2.40)

and the overall mean free path for the mixture is

$$\lambda_0 = \sum_{p=1}^{s}\left\{\left(n_p/n\right)\left(\lambda_p\right)_0\right\}.$$

(2.41)

2.4 The Chapman-Enskog transport properties

As noted above, the reference diameter of a molecular species is set to the value that reproduces the coefficient of viscosity at the reference temperature. The Chapman-Enskog theory (Chapman and Cowling, 1970) provides the necessary relationships. This theory provides a solution of the Boltzmann equation that is based on the assumption that the distribution function is a small perturbation of the equilibrium distribution. *i.e.*

$$f = f_0\left(1+\Phi\right)$$

where Φ is small in comparison with unity.

The Chapman-Enskog result for the coefficient of viscosity in a VHS gas is (Bird, 1994)

$$\mu = \frac{15\left(\pi mk\right)^{1/2}\left(4k/m\right)^{\omega-1/2}T^{\omega}}{8\Gamma\left(\tfrac{9}{2}-\omega\right)\sigma c_r^{2\omega-1}}.$$

(2.42)

This is an approximate result in that the perturbation Φ can be regarded as the first order term of a power law expansion in the local inverse Knudsen number. Moreover, Chapman's solution involves a series of Sonine polynomials and Eqn. (2.42), which is based on the first order term in this expansion, is called the "first approximation".

For a gas that has a coefficient of viscosity proportional to T^ω and with the value μ at temperature T, Eqns. (2.28), (2.32), and (2.42) show that the diameter is

$$d = \left\{ \frac{15(mkT/\pi)^{1/2}}{2(5-2\omega)(7-2\omega)\mu} \right\}^{1/2}.$$

(2.43)

The first approximation to the coefficient of heat conduction K is related to the coefficient of viscosity by

$$K = {}^{15}\!/_{\!4}\, k\mu / m,$$

(2.44)

so that the Prandtl number in a monatomic gas, for which the specific heat at constant pressure c_p is equal to $\frac{5}{2}k/m$, is

$$P_r \equiv \mu c_p / K = \tfrac{2}{3}.$$

(2.45)

This has been found to be a realistic prediction.

The first approximation to the diffusion coefficient for species 1 and 2 is

$$D_{12} = \frac{3\pi^{1/2}(2kT/m_r)^{\omega_{12}}}{8\,\Gamma(\tfrac{7}{2}-\omega_{12})\,n\,\sigma_{12}\,c_r^{2\omega_{12}-1}}$$

(2.46)

and the coefficient of self-diffusion is

$$D_{11} = \frac{3\pi^{1/2}(4kT/m)^{\omega}}{8\,\Gamma(\tfrac{7}{2}-\omega)\,n\,\sigma\,c_r^{2\omega-1}}.$$

(2.47)

Comparing Eqns. (2.42) and (2.47), the Schmidt number for the VHS model is

$$S_c \equiv \mu/(\rho D_{11}) = 5/(7-2\omega).$$

(2.48)

Equation (2.47) can be combined with Eqns. (2.28) and (2.32) to obtain a molecular diameter based on the self-diffusion coefficient. This is

$$d = \left\{ \frac{3(mkT/\pi)^{1/2}}{2(5-2\omega)\rho D_{11}} \right\}^{1/2}$$

(2.49)

and is, of course, equal to the diameter of Eqn. (2.43) when the Schmidt number is given by Eqn. (2.48). This equation predicts a

Schmidt number of $^{10}/_{11}$ for the value of $\omega = 0.75$ which is typical of the measured viscosity-temperature power laws in real gases. The Schmidt number at the hard sphere limit of a VHS gas is $^5/_6$, while the Maxwell gas value is unity. The Chapman-Enskog predictions of the diffusion and viscosity coefficients in an inverse power law gas also predict a Schmidt number of $^5/_6$ for a hard sphere gas, but the predicted value for a Maxwell gas is smaller rather than larger at 0.645. The measured values of the self-diffusion coefficient in real gases lead to Schmidt numbers of the order of 0.75 and are therefore consistent with the predictions of the inverse power law models. The unrealistic prediction of the Schmidt number by the VHS model has led to the development of the VSS model that can reproduce the measured Schmidt numbers. The VSS theory is presented in § 3.3.

2.5 Gas-surface interactions

The classical models for gas-surface interactions date back to the earliest days of the kinetic theory of gases. Maxwell (1879) proposed two models for the interaction of an equilibrium gas with a solid surface that maintain equilibrium. **Specular reflection** is perfectly elastic with the molecular velocity component normal to the surface being reversed, while the velocity components parallel to the surface remain unaffected. In **diffuse reflection** the velocity of each molecule after reflection is independent of its velocity before reflection. However, the velocities of the reflected molecules as a whole are distributed in accordance with the half-range Maxwellian or equilibrium for the molecules that are directed away from the surface. Equilibrium diffuse reflection requires that both the surface temperature and the temperature associated with the reflected gas Maxwellian be equal to the gas temperature. In the case of specular reflection, the gas may have a stream velocity parallel to the surface and a specularly reflecting surface is functionally identical to a **plane of symmetry**.

The general requirement, at the molecular level, for equilibrium between a solid surface and a gas is that the interaction should satisfy the **reciprocity condition**. This is a relationship between the probability of a gas-surface interaction with a particular set of incident and reflected velocities and the probability of the inverse interaction. It may be written (Cercignani, 1969) as

$$c_r.e\,P(-c_r,-c_i)\exp\left(\frac{-E_r}{kT_W}\right) = -c_i.e\,P(c_i,c_r)\exp\left(\frac{-E_i}{kT_W}\right) \qquad (2.50)$$

The unit vector e has been taken normal to the surface which is at temperature T_W. $P(c_1,c_2)$ is the probability that a molecule incident on a surface with velocity c_1 leaves with velocity c_2, and E is the energy of the molecule. This condition is related to the law of detailed balance and is satisfied by both the diffuse and specular models for a gas in equilibrium with a surface. While most DSMC applications deal with nonequilibrium situations, the procedures for gas-surface interactions must be such that reciprocity is satisfied when they are applied to equilibrium situations.

A gas generally has a velocity component parallel to a surface and this means that the stagnation temperature in the gas differs from the static temperature. For other than fully specular reflection, the distribution function for the incoming molecules will be different to that for the reflected molecules and the distribution function for the molecules near the surface will not be a Maxwellian. Also, the energy of a molecule relative to the surface before it strikes the surface will generally be different from the energy relative to the surface after it has been reflected from the surface, so that the process is inelastic. It has been found that, for the surfaces that are encountered in engineering problems, there is generally good agreement with calculations that assume that the reflected molecules are diffusely reflected at the surface temperature. This means that the reflected molecules are described by the equations of §2.2 in that they are the flux from a fictitious equilibrium gas on the opposite side of the surface. The temperature of this gas is equal to the surface temperature and its density is such that the number flux of Eqn. (2.12) is equal to the number flux that is incident on the surface.

There are significant departures from diffuse reflection when a surface, particularly a heated surface, has been exposed for a long period to ultra-high vacuum. In the case of very light molecules incident on a smooth clean surface of a heavy metal, there is evidence that the reflection may be near specular. There were attempts to deal with this problem through the definition of momentum and energy accommodation coefficients. These were defined in different ways and it was found that many were inconsistent with the reciprocity principle. However, the major problem is that the results are more sensitive to the contaminants on and the history of the surface than to its composition. This is the greatest problem associated with DSMC predictions for engineering problems that involve surfaces that have been exposed to high vacuum in space or in vacuum systems. This means that it is impossible to confidently predict a precise result and the likely limits of the uncertainty must be established through sensitivity studies. These are most often made by setting a combination of diffuse and specular reflection. However, molecular beam experiments indicate that the scattering is in a lobe about the specular direction rather than in the specular direction.

Cercignani and Lampis (1974) produced a gas surface interaction model that not only satisfies the reciprocity condition, but can match the measured scattering profiles (Woronowicz and Rault, 1994) if information is available for several accommodation coefficients. Its implementation in DSMC codes has been facilitated by a graphical interpretation of the model by Lord (1991) and it is generally called the CLL model.

The model assumes that there is no coupling between the normal and tangential components of the velocity during the reflection process. Set u to be the normal component of the molecular velocity normalized to the most probable molecular speed at the surface temperature, and v and w to be the similarly normalized tangential components. The theory employs tangential and normal accommodation coefficients α_t and α_n. The probability that an incident molecule with a normalized tangential component v_i is reflected with tangential component v_r is

$$P(v_i, v_r) = \{\pi \alpha_t\}^{-1/2} \exp\left[-\left\{v_r - (1-\alpha_t)^{1/2} v_i\right\}^2 \Big/ \alpha_t\right] \qquad (2.51)$$

with a similar equation for w. The corresponding probability for the normal component is

$$P(u_i, u_r) = (2u_r / \alpha_n) I_0 \left\{2(1-\alpha_n)^{1/2} u_i u_r / \alpha_n\right\} \exp\left[\left\{u_r^2 + (1-\alpha_n)u_i^2\right\} \big/ \alpha_n\right]. \qquad (2.52)$$

It can be shown that the probability function for $(v^2 + w^2)^{1/2}$ also satisfies this equation, as does the equation for the two components of angular velocity of a diatomic molecule.

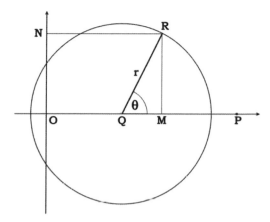

Fig. 2.3 Lord's geometrical representation of the Cercignani-Lampis model.

Lord (1991) introduced the graphical construction of Fig. (2.3) to illustrate this general probability function. The state of the incident molecule is represented by the point P on the horizontal axis and its distance OP from the origin can represent the magnitude of any one of u_i, $\left(v_i^2 + w_i^2\right)^{1/2}$, or ω_i. The point Q represents the average state of the reflected molecule and is situated on OP such that $OQ/OP = \left(1 - \alpha\right)^{1/2}$, where α is the appropriate energy accommodation factor. The actual state of the reflected molecule is represented by the point R and the probability distribution of this state is given by a two-dimensional Gaussian centred on Q. This shows that the probability of the state lying in the element $r\,dr\,d\theta$ at $\left(r,\theta\right)$ is

$$\left(\pi\alpha\right)^{-1}\exp\left(-r^2/\alpha\right)r\,dr\,d\theta. \tag{2.53}$$

where r is the distance QR and θ is the angle PQR. Note that θ is uniformly distributed from 0 to π. The distance OR represents u_r, $\left(v_r^2 + w_r^2\right)^{1/2}$, or ω_r, while the projections OM and ON onto the axes represent either v_r and w_r or the two components of the angular velocity of a diatomic molecule.

References

Bird, G. A. (1994). *Molecular Gas Dynamics and the Direct Simulation of Gas Flows*, Clarendon Press, Oxford.

Cercignani, C. (1969). *Mathematical Methods in Kinetic Theory*, Plenum Press, New York.

Cercignani, C. and Lampis, M. (1974). In *Rarefied Gas Dynamics*, (Ed. K. Karamcheti), p. 361, Academic Press, New York.

Chapman, S. and Cowling, T.G. (1970). *The Mathematical Theory of Non-uniform Gases* 3rd edn., Cambridge University Press.

Lord, R. G. (1991). Some extensions to the Cercignani-Lampis gas scattering kernel, *Phys. Fluids A* **3**, 706-710.

Maxwell, J. C. (1879). Phil. Trans. Roy. Soc. 1, Appendix.

Woronowicz, M. S. and Rault, D. F. G. (1994). Cercignani-Lampis-Lord gas-surface interaction model- Comparisons between theory and simulation, *J. Spacecraft and Rockets* **31**, 532-534.

3

MOLECULAR MODELS FOR DSMC

3.1 Phenomenological modeling

As shown in Figure 3.1, early applications of DSMC employed the molecular models that had been employed in classical kinetic theory. The **point centre of force** model and the **Lennard-Jones** model (that adds a short-range attractive force to the long-range repulsive force) can be classified as **approximate physical models** and remain useful for DSMC. The **Maxwell molecule** is a **mathematical model** that was introduced to simplify the theory. The much-used **hard sphere** model for elastic collisions can be classed as a **mechanical analog**. All the classical models for molecules with internal degrees of freedom also fall into this class. These provide a poor representation of a real gas and are difficult to implement as DSMC models. They have been largely replaced by **phenomenological molecular models** that merely aim to reproduce the observed properties of real gases.

With regard to the classical kinetic theory models for monatomic gases, the point centre of force model assumes an inverse repulsive power law and the hard sphere molecule can be regarded as the limiting "hard" case in which the power law tends to infinity, while the Maxwell molecule is the special "soft" case in which the force is proportional to the inverse fifth power. The hard sphere model is the most easily applied in that it has a constant cross-section. It is also the most computationally efficient because the scattering is isotropic in the centre of mass frame of reference. On the other hand, the inverse power law extends indefinitely and an arbitrary cut-off in diameter or deflection angle must be imposed when it is employed in DSMC. Moreover, the expression for the deflection angle is so complex that its analytical evaluation at every collision is out of the question and its implementation requires a single parameter look-up table. Nevertheless, it was employed in early DSMC studies including a study (Bird, 1970) of the effect of molecular model on the structure of a normal shock wave.

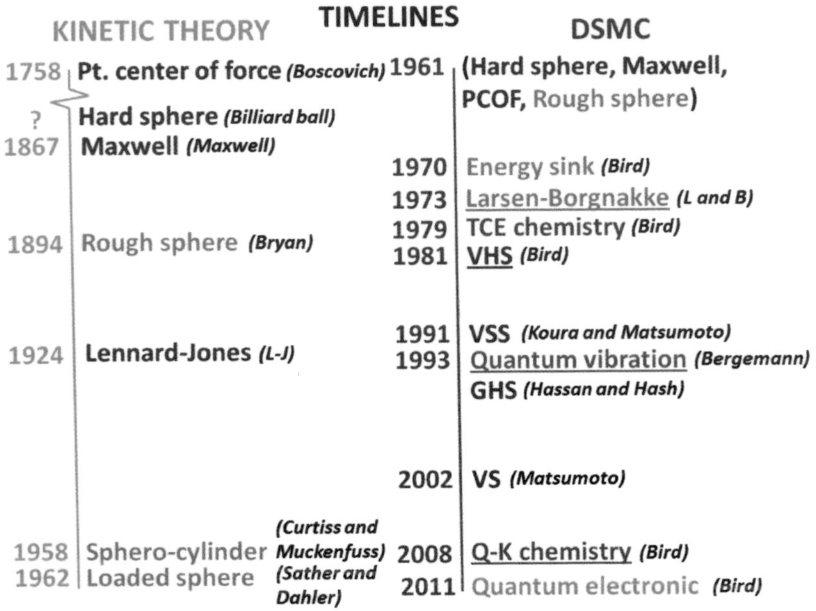

Fig. 3.1 The transition from mechanical analogs to phenomenological models.

The early choice of model was unduly influenced by the discussions in the classical kinetic theory texts (for example, Chapman and Cowling, 1970) that concentrate almost entirely on the distribution of the deflection angle in collisions. The effect of the temperature on the collision cross-section is implicit in the analysis, but is not mentioned in the discussion. The shock wave study was one of many that eventually led to the realization that, for all the flows that had been studied, the effects of molecular model correlate almost exactly with the change in collision cross-section and that the way in which the molecules were scattered in collisions was apparently of little consequence. This led (Bird, 1981) to the variable hard sphere, or VHS model, that combines the simplicity of the hard sphere model with the improved accuracy of the inverse power law model that includes the temperature dependence of the coefficient of viscosity. Its implementation in a DSMC program is incomparably simpler than that of the inverse power law model. It has the further advantage over the inverse power law model in that there is no need for an arbitrary diameter or deflection angle cut-off, so that the mean free path and mean collision rate can be unambiguously defined.

The power law of repulsion η in the point centre of force model is generally about 12 and is related to the temperature-viscosity index by

$$\omega = \tfrac{1}{2}(\eta + 3) / (\eta - 1). \tag{3.1}$$

The Lennard-Jones, or L-J, model adds a soft point centre of attraction, generally of power 6, to the hard repulsion and this component becomes significant at very low temperatures. However, it requires a two parameter look-up table for its implementation in a DSMC program and this is justified only if it can be shown that the additional effort is necessary.

The VHS model can be regarded as a phenomenological model in that the attainment of realistic transport properties at the macroscopic level is given priority over the employment of more realistic molecular potentials at the microscopic level. The introduction of the VHS model for elastic collisions came some years after the introduction by Larsen and Borgnakke (1974) of their model for rotation in diatomic and polyatomic molecules.

The Larsen-Borgnakke model can be regarded as the archetypal phenomenological model in that it sought only to maintain the equilibrium rotational distribution function while reproducing a specified rotational relaxation collision number. It is interesting to note that the Larsen-Borgnakke distributions were introduced in the context of the DSMC method almost simultaneously with the introduction by Levine and Bernstein (1974) of the equivalent "prior" distributions in the chemistry literature. While the Larsen-Borgnakke distributions have since been generalized (Bird, 1994) to consistently account for the effects of molecular model, the prior distributions remain (Levine and Bernstein, 2005) in their original form. The translational term in the prior distributions is valid only for Maxwell molecules, while the vibrational term is valid only for hard sphere molecules. The extension to vibration is complicated by the fact that vibration is generally only partially excited and the classical extension of the Larsen-Borgnakke approach to this mode was computationally inefficient and dealt only with the overall vibrational energy. The situation was transformed by the introduction by Bergemann (1994) of the quantum model for vibration. This stored the vibrational level of each simulated molecule instead of the vibrational energy and enormously simplified the DSMC procedures. The harmonic oscillator model for vibration is generally employed, but the extension of the approach to the unevenly spaced electronic modes demonstrates that the assumption of equally spaced levels is not necessary. Quantum procedures can be applied also to the rotational mode when the simulation includes effects that depend on the rotational state.

The quantum treatment of vibration permits a physically realistic treatment of dissociation in that it occurs in a collision whenever the vibrational excitation is to the level that corresponds to dissociation. This has been implemented in the Quantum-Kinetic or Q-K model (Bird, 2011). The procedure for a recombination reaction is to implement it whenever there is a probability of another molecule being within the "ternary collision volume." This volume is set to the value that satisfies the equilibrium constant for the forward dissociation reaction and the reverse recombination reaction. The procedures for binary exchange and chain reaction are analogous to the dissociation procedure, but are entirely phenomenological. Most applications of DSMC to chemically reacting flows have employed procedures (Bird, 1979) that are now called the TCE method. These convert experimentally based rate coefficients to reaction cross-sections. The Q-K procedures are sufficiently simple that analytical expressions can be derived for the corresponding rate coefficients in an equilibrium gas. The Q-K predicted rates have been shown (e.g., Gallis et al, 2010) to be generally within the bounds of uncertainty of the generally accepted rates. Moreover, because the predicted rates are in the form of a summation over the vibrational levels, expressions may be written down for the vibrationally resolved rates. These allow the post-reaction vibrational state of the product molecules to be set to values that satisfy the detailed balance principle. The Larsen-Borgnakke procedures lead to equilibrium in a non-reacting gas, but their application to reacting gases, as in the TCE model, leads to a failure of equipartition with regard to the vibrational temperature.

A typical DSMC application involves millions of simulated molecules and trillions of intermolecular collisions. The best possible solution of the Schrödinger equation for single collision poses a computational task of a similar magnitude to that of the whole DSMC computation. It is therefore necessary to either employ the phenomenological models or refer to a vast database of the results from more physically realistic, but unavoidably approximate, computations that cover all collisional impact parameters for all molecular species. The vast database does not exist and probably never will but, as noted earlier, a single parameter look-up table was employed in the early implementation of the inverse power law model. This was superseded by the VHS model that achieved only a small gain in computational efficiency due largely to the isotropic scattering, but was incomparably easier to implement. As noted earlier, the L-J model requires a two-parameter look-up table and even more effort is needed for its implementation. Nevertheless, it has been implemented by Sharipov and Strapasson (2012) who claimed that *"....the proposed scheme dispenses the widely used variable hard-sphere, variable soft sphere, and any other similar model."* The startling recommendation to dispense with the VHS model was apparently based on the assumption that a model that has a greater degree of physical realism necessarily provides a more accurate result. The additional effort and the disadvantages associated with the arbitrary cut-off can be justified only if the L-J model leads to results that are demonstrably superior to the VHS results. Results from the L-J model were provided for several test cases and this has permitted a direct comparison with VHS results from the unmodified public domain program **DS1.EXE**.

The benchmark case is for the one-dimensional heat flux in helium between a surface at 350 K and a parallel surface at 250 K that is separated by 1 mm from the first surface.

Table 3.1 A comparison of the heat flux in kW/m² for Lennard-Jones molecules with that for the VHS and hard sphere models.

Pressure (Pa)	Lennard-Jones	VHS	Hard sphere
100	8.415	8.425	8.503
200	10.93	10.94	11.03
400	12.90	12.88	12.95

Table 3.1 shows that the results for the VHS model are in agreement with the L-J results to within 0.2% and that those for the hard sphere model are only 1% higher. The soft attractive component is expected to be significant at sufficiently low temperatures and the GHS model then provides an alternative to the L-J model, but there are no observations on which to base comparative tests. There is no case for dispensing with the VHS and VSS models at higher temperatures.

All molecular models are phenomenological to some extent and, while the degree of physical realism should be kept in mind, the accuracy of the results that are produced by the model is the primary consideration. **A model with a greater degree of physical realism does not necessarily lead to a more accurate result!**

3.2 Binary collision mechanics

The models for internal energy modes are add-ons to the VHS and VSS models for the translational modes. They affect the translational modes only to the extent that the post-collision translational energy differs from the pre-collision translational energy. The possibility of chemical reactions must be taken into account and there may be changes in the pre and post-collision molecular masses.

Consider a collision between two molecules of mass m_1 and m_2 with velocity components $\boldsymbol{c}_1$ and $\boldsymbol{c}_2$, respectively. The corresponding post-collision variables are m_1^*, m_2^*, $\boldsymbol{c}_1^*$ and $\boldsymbol{c}_2^*$. Conservation of mass and momentum requires that

$$m_1 + m_2 = m_1^* + m_2^* \tag{3.2}$$

and

$$m_1\boldsymbol{c}_1 + m_2\boldsymbol{c}_2 = m_1^*\boldsymbol{c}_1^* + m_2^*\boldsymbol{c}_2^*.$$

(3.3)

The **centre of mass velocity** $\boldsymbol{c}_m$ that is defined by

$$\boldsymbol{c}_m = \frac{m_1\boldsymbol{c}_1 + m_2\boldsymbol{c}_2}{m_1 + m_2} = \frac{m_1^*\boldsymbol{c}_1^* + m_2^*\boldsymbol{c}_2^*}{m_1^* + m_2^*} \tag{3.4}$$

is therefore unchanged in the collision. The pre-collision and post-collision relative velocities of the molecules are

$$\boldsymbol{c}_r = \boldsymbol{c}_1 - \boldsymbol{c}_2 \quad \text{and} \quad \boldsymbol{c}_r^* = \boldsymbol{c}_1^* - \boldsymbol{c}_2^*. \tag{3.5}$$

Equations (3.4) and (3.5) may be combined to obtain expressions for the molecular velocities in terms of the relative and centre of mass velocities. Those for the pre-collision velocities are

$$\boldsymbol{c}_1 = \boldsymbol{c}_m + \frac{m_2}{m_1 + m_2}\boldsymbol{c}_r \quad \text{and} \quad \boldsymbol{c}_2 = \boldsymbol{c}_m - \frac{m_1}{m_1 + m_2}\boldsymbol{c}_r \tag{3.6}$$

and those for the post collision velocities are

$$c_1^* = c_m + \frac{m_2^*}{m_1^* + m_2^*} c_r^* \quad \text{and} \quad c_2^* = c_m - \frac{m_1^*}{m_1^* + m_2^*} c_r^*. \tag{3.7}$$

When considering the translational energies associated with the collision, it is customary to introduce the reduced mass $m_r = m_1 m_2 / (m_1 + m_2)$ that was defined in Eqn. (2.24). The pre-collision translational energy of the molecules can then be written down from Eqns. (3.6) and (3.7) as

$$E_{tr} = \tfrac{1}{2} m_1 c_1^2 + \tfrac{1}{2} m_2 c_2^2 = \tfrac{1}{2}(m_1 + m_2) c_m^2 + \tfrac{1}{2} m_r c_r^2. \tag{3.8}$$

Mass is conserved in the collision and the centre of mass velocity is unchanged so that the corresponding post-collision translational energy is

$$E_{tr}^* = \tfrac{1}{2} m_1^* c_1^{*2} + \tfrac{1}{2} m_2^* c_2^{*2} = \tfrac{1}{2}(m_1 + m_2) c_m^2 + \tfrac{1}{2} m_r^* c_r^{*2}. \tag{3.9}$$

It should be noted that, for collisions that involve chemical reactions, the post-collision reduced mass differs from the pre-collision value.

The initial data for the DSMC computation of a binary collision comprises the molecular species, the three velocity components of each molecule and either the energy in or the quantum state of each internal mode. The collision routine involves the following steps:

(i)   The components of the centre of mass and the relative velocity vectors are calculated from Eqns. (3.4) and (3.5) and the relative translational energy in the collision $\tfrac{1}{2} m_r c_r^2$ is calculated.

(ii) The total collision energy is the sum of the relative translational energy and the energies in all the internal modes of both molecules.

(iii) In the case of chemically reacting flows, the heats of formation of each molecule are added to the total collision energy. The reaction model is then applied and, if the reaction occurs, the molecular species are set to the post-reaction identities and the total collision energy is adjusted for the changed heats of formation.

(iv) The models for the internal modes are applied and the pre-collision energies are either retained or adjusted. The post-collision internal energies are subtracted from the total collision energy to obtain the post-collision relative translational energy $\tfrac{1}{2} m_r^* c_r^{*2}$.

(v) The magnitude of the post-collision relative velocity is calculated and the elastic collision model is applied, as described in Section 3.3, to determine the components of the post-collision relative velocity. Equation (3.7) is then used to calculate the post-collision velocity components in the laboratory frame of reference.

3.3 Elastic collision models

The variable hard sphere, or VHS, model has already been discussed. Most DSMC applications employ this model and the VSS model is employed only when it is necessary to reproduce the Schmidt number in diffusing gas mixtures. The cross-sections of both the VHS and VSS models are chosen to match the coefficient of viscosity at some reference temperature and the dependence of the cross-sections on the relative translational energy is chosen to match the dependence of the viscosity on temperature in the real gas. The VSS model involves an empirical modification of the isotropic scattering law and the basic hard sphere collision mechanics that are illustrated in Fig. (3.2) apply to both the VHS and VSS model.

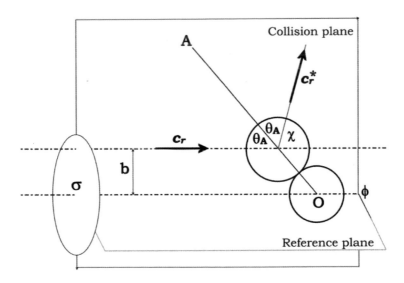

Fig. 3.2 The collision mechanics of hard sphere molecules in the centre of mass frame of reference.

A collision is planar in the centre of mass frame of reference and the **impact parameters** are the angle ϕ between the collision plane and a plane of reference and the miss distance b between lines through the centres of the molecules that are parallel to the pre-collision relative velocity. As shown in Fig. 3.2, the origin O is at the centre of one of the molecules and the line OA through the centre of the other molecule defines the **apse line**. The apse line bisects the angle between the pre-collision and post collision relative and the deflection angle is defined as

$$\chi = \pi - 2\theta_A,$$ (3.10)

and $\chi/2$ is complementary to θ_A. The geometry in Fig. 3.2 is such that

$$\chi = 2\cos^{-1}(b/d_{12}),$$ (3.11)

where $d_{12} = (d_1 + d_2)/2$ is the mean diameter of the two molecules. The element of solid angle into which the molecule is scattered is

$$d(\Omega) = \sin\chi\, d\chi\, d\phi = (4/d_{12}^2)b\, db\, d\phi = (4/d_{12}^2)d(\sigma).$$ (3.12)

The element of solid angle is therefore related to the element of cross-section by a constant so that the hard sphere scattering is isotropic in the centre of mass frame of reference. The total solid angle is 4π so that the total collision cross-section for hard sphere molecules is, as expected,

$$\sigma = \pi d_{12}^2.$$ (3.13)

The direction of the post-collision relative velocity is set by randomly selected azimuth and elevation angles. *i.e.*

$$\cos\theta = 2R_F - 1, \quad \sin\theta = \sqrt{1 - \cos^2\theta}$$

and (3.14)

$$\phi = 2\pi R_F.$$

The three components of post-collision relative velocity are therefore

$$u_r^* = \cos\theta c_r^*, \quad v_r^* = \sin\theta\cos\phi c_r^* \quad \text{and} \quad w_r^* = \sin\theta\sin\phi c_r^* \quad (3.15)$$

The VSS model replaces Eqn. (3.11) with

$$\chi = 2\cos^{-1}\left\{(b/d_{12})^{1/\alpha}\right\}, \qquad (3.16)$$

where α is the empirical factor that is chosen to match the Schmidt number in flows that are dominated by diffusion. Because the scattering is no longer isotropic, the generation of the post-collision relative velocity components is far more complex. From Eqn. (3.16), a representative set of values for the cosine and sine of the deflection angle is given by

$$\cos\chi = 2R_F^{1/\alpha} - 1, \quad \text{and} \quad \sin\chi = \sqrt{1-\cos^2\chi}. \qquad (3.17)$$

Now introduce a set of Cartesian coordinates x', y' and z' with x' in the direction of $\mathbf{c}_r$. The components of $\mathbf{c}_r^*$ along these axes are

$$c_r^* \cos\chi, \quad c_r^* \sin\chi\cos\phi \quad \text{and} \quad c_r^* \sin\chi\sin\phi.$$

The direction cosines of x' are

$$u_r/c_r, \quad v_r/c_r \quad \text{and} \quad w_r/c_r.$$

Since the orientation of the reference plane is arbitrary, the y' axis may be chosen such that it is normal to the x-axis. The direction cosines of y' and z' are then

$$0, \quad w_r/\sqrt{v_r^2+w_r^2} \quad \text{and} \quad -v_r/\sqrt{v_r^2+w_r^2}$$

and

$$\sqrt{v_r^2+w_r^2}/c_r, \quad -u_r v_r/\left(\sqrt{v_r^2+w_r^2}\,c_r\right) \quad \text{and} \quad -u_r w_r/\left(\sqrt{v_r^2+w_r^2}\,c_r\right).$$

The required expressions for the components of $\mathbf{c}_r^*$ in the original x, y and z coordinates are

$$u_r^* = \left[\cos\chi\, u_r + \sin\chi\sin\phi\sqrt{v_r^2+w_r^2}\right]c_r^*/c_r,$$

$$v_r^* = \left[\cos\chi\, v_r + \sin\chi\,(c_r w_r \cos\phi - u_r v_r \sin\phi)/\sqrt{v_r^2+w_r^2}\right]c_r^*/c_r \quad (3.18)$$

and $\quad w_r^* = \left[\cos\chi\, w_r - \sin\chi\,(c_r v_r \cos\phi + u_r w_r \sin\phi)/\sqrt{v_r^2+w_r^2}\right]c_r^*/c_r.$

The Chapman-Enskog theory has been applied (Bird, 1994) to the VSS model. The VSS equations that correspond to the VHS Eqns. (2.43) and (2.49) for the molecular diameters based the viscosity and self-diffusion coefficients are

$$d = \left\{ \frac{5(\alpha+1)(\alpha+2)(mkT/\pi)^{1/2}}{4\alpha(5-2\omega)(7-2\omega)\mu} \right\}^{1/2} \tag{3.19}$$

and

$$d = \left\{ \frac{3(\alpha+1)(mkT/\pi)^{1/2}}{4(5-2\omega)\rho D_{11}} \right\}^{1/2}. \tag{3.20}$$

The two diameters are equal if α is related to the Schmidt number by

$$\alpha = \frac{10}{S_c(21-6\omega)-5}. \tag{3.21}$$

3.4 Models for internal energy

The Larsen-Borgnakke, or L-B, procedures set the distribution of post-reaction internal energies to values that satisfy the detailed balance principle. The procedures therefore maintain the equilibrium distribution of all modes and the establishment of equilibrium for each mode automatically leads to equilibrium between modes and to equipartition. It is also necessary to reproduce the observed relaxation rates when the gas is initially not in equilibrium. The original version of the method dealt primarily with the rotational mode and, in order to match a rotational relaxation collision number Z_{rot}, the rotational energy adjustment procedure was applied to one collision in every Z_{rot} collisions. However, every collision in a real diatomic or polyatomic gas involves changes to the rotational energies and, in order to increase the degree of physical realism, Larsen and Borgnakke introduced the "restricted energy exchange" modification of their method such that $1/Z_{rot}$ of the adjustment was made in every collision. Not only did this make the procedures inefficient from the computational point of view, but it was shown (Pullin, 1978) that the modified procedures do not satisfy detailed balance. The method is therefore applied in its original form and it can be shown that the fractional application procedure leads to the desired relaxation rate.

Separate L-B procedures are required for each internal energy mode and an energy adjustment for any mode involves a redistribution of energy between modes. There would therefore be a difficulty in its implementation if the post-collision distribution of energy was dependent on the order in which the procedures are applied to the two molecules in the collision or to the internal modes of each molecule. This question has been investigated analytically (Bird, 1964, §5.5) and it was shown that the final distribution of energies is independent of the order of application of the procedures. The L-B process is therefore unaffected if there is a succession of redistributions, each of which involves only a single internal energy mode of one molecule and the translational mode that is, in the centre of mass frame of reference, common to the molecules. The energy that is available for redistribution is this relative translational energy plus the pre-collision energy in the mode under consideration. Although the available energy decreases as the successive redistributions are made, this is exactly balanced by a higher probability of selection of a particular energy of the internal mode under consideration.

A general L-B distribution function for the division of post-collision energy between the energy modes has been developed (Bird, 1964, §5.5). Consider a collision between a molecule of species 1 and a molecule of species 2. The parameter Ψ is the average over the two molecules of the sum of the degrees of freedom. Equation (2.31) shows that the number of degrees of freedom associated with the relative translational energy is $5 - 2\omega_{12}$ and the average is half this number. Therefore

$$\Psi = 5/2 - \omega_{12} + \zeta_{rot,1}/2 + \zeta_{rot,2}/2 + \zeta_{vib,1}/2 + \zeta_{vib,2}/2 . \qquad (3.22)$$

Let Ψ_a be one or more of the terms on the right hand side of Eqn. (3.22) and let Ψ_b be the remaining terms that are participating in the partitioning of energy. The energy that is assigned to the first mode or groups of mode is E_a and E_b is that to be assigned to the second. The sum of these energies is called the collision energy $E_c = E_a + E_b$ even though it generally does not include the energy in all the modes of the two molecules. The collision energy is a known quantity and is a constant in the energy redistribution process. The general L-B result for the distribution of energy between the two groups is then

$$f\left(\frac{E_a}{E_c}\right) = f\left(\frac{E_b}{E_c}\right) = \frac{\Gamma(\Psi_a + \Psi_b)}{\Gamma(\Psi_a)\Gamma(\Psi_b)}\left(\frac{E_a}{E_c}\right)^{\Psi_a - 1}\left(\frac{E_b}{E_c}\right)^{\Psi_b - 1}. \qquad (3.23)$$

This equation provides the post-collision distribution function for both E_a and E_b. DSMC implementations take advantage of the order independence that was discussed in the preceding paragraph and make a series of redistributions between the translational energy and a single internal mode. Ψ_a is then half the number of degrees of freedom in the internal mode and Ψ_b is $5/2 - \omega_{12}$. The distribution function for the ratio of the post-collision internal energy E_a of the mode to the sum E_c of the internal and translational energy, that is a constant in the redistribution, is

$$f\left(\frac{E_a}{E_c}\right) = \frac{\Gamma(\zeta/2 + 5/2 - \omega_{12})}{\Gamma(\zeta/2)\Gamma(5/2 - \omega_{12})}\left(\frac{E_a}{E_c}\right)^{\zeta/2 - 1}\left(1 - \frac{E_a}{E_c}\right)^{3/2 - \omega_{12}}. \qquad (3.24)$$

For example, consider the special case of the assignment of energy to a single internal mode with two degrees of freedom. Eqn. (3.24) then reduces to

$$f\left(\frac{E_a}{E_c}\right) = (5/2 - \omega_{12})\left(1 - \frac{E_a}{E_c}\right)^{3/2 - \omega_{12}}. \qquad (3.25)$$

This distribution is amenable to sampling from the cumulative distribution function and the representative value of E_a that corresponds to the random fraction R_F is

$$E_a = E_c\left(1 - R_F^{1/(5/2 - \omega_{12})}\right). \qquad (3.26)$$

This can be applied directly to the rotational energy of a diatomic molecule. The case for non-linear polyatomic molecules with three degrees of freedom requires an acceptance-rejection procedure based on the ratio of the probability of a particular value of E_a/E_c to the maximum probability. There is a maximum probability of P_u/P_c in the range of interest from 0 to 1 when $\zeta > 2$. This is

$$\frac{P}{P_{max}} = \left\{\frac{\zeta/2 + 1/2 - \omega_{12}}{\zeta/2 - 1}\left(\frac{E_a}{E_c}\right)\right\}^{\zeta/2 - 1}\left\{\frac{\zeta/2 + 1/2 - \omega_{12}}{3/2 - \omega_{12}}\left(1 - \frac{E_a}{E_c}\right)\right\}^{3/2 - \omega_{12}}. \qquad (3.27)$$

The preceding analysis has implicitly assumed that the energies in the modes are continuously distributed. This is a valid assumption for translation because of the enormous number of available states and is a good approximation for the rotational mode at normal or higher temperatures. A **quantum model** is required for the **vibrational and electronic** modes. A diatomic gas has only one vibrational mode, while polyatomic molecules generally have more than one mode. While monatomic molecules have no rotational or vibrational modes, they do have electronic modes

Consider the quantum case in which the energy in a particular internal mode of a molecule is restricted to that of the discrete states ε_i where $i = 0$ to j. The zero point energy has no effect and can be ignored, so that $\varepsilon_0 = 0$. Now consider the L-B redistribution between the translational mode as group b and a quantized internal mode with two degrees of freedom of molecule 1 as group a. Equation (3.25) can then be applied with the energy E_c equal to the sum of the pre-collision translational energy and the energy of the pre-collision state. E_c is a constant in the redistribution and the probability of post-collision state i^* is proportional to $\left(1 - \varepsilon_{i^*} / E_c\right)^{3/2 - \omega_{12}}$. The maximum probability is that of the ground state, so that the ratio of the probability to the maximum probability is

$$\frac{P}{P_{max}} = \left(1 - \frac{\varepsilon_{i^*}}{E_c}\right)^{3/2 - \omega_{12}}. \tag{3.28}$$

The acceptance-rejection procedure may be applied to this probability ratio in order to select the post-collision state of molecule 1. This selection process is applied to potential states that are chosen uniformly from the ground state to the highest state with energy below E_c. Note that the selection procedure involves states rather than levels and the degeneracy g_k of a level k must be taken into account.

The spacing of the energy levels is not uniform but, in the case of vibration, most applications employ the **simple harmonic model** that assumes uniformly spaced levels

$$\varepsilon_{vib,i} = i k \Theta_{vib} \tag{3.29}$$

where Θ_{vib} is the characteristic vibrational temperature. Equation (3.28) then becomes

$$\frac{P}{P_{max}} = \left(1 - \frac{i^* k \Theta_{vib}}{E_c}\right)^{3/2 - \omega_{12}}.$$ (3.30)

3.5 Relaxation rates

As noted above, if the collision number of a relaxing mode is Z, the collision procedure for a particular mode is applied in $1/Z$ of the relevant collisions. The relaxation collision numbers are set as part of the data for each molecular species. These collision numbers are often an arbitrary function of the temperature and the question arises as to how this temperature dependence should be applied within the DSMC procedures.

One option is to employ the sampled macroscopic temperature. This is straightforward to implement and will always lead to the desired relaxation rate. On the other hand, there is an argument that the molecules in a collision are unaware of the macroscopic temperature and that the collision procedure should be based entirely on the microscopic information associated with the colliding molecules. The "collision-based" procedures generally involve the definition of a "collision temperature" and the ambiguity associated with this definition has led to a number of questionable procedures.

To the extent that a collision temperature will be required, the problems associated with potentially non-equilibrium modal distributions are avoided if it is defined through the translational energy alone. Equation (2.31) can be written

$$E_{tr} = \tfrac{1}{2} m_r c_r^2 = \left(\tfrac{5}{2} - \omega\right) k T$$

and the definition of the collision temperature based only on the relative translational energy in the collision follows as

$$T_{coll} = \frac{m_r c_r^2}{(5 - 2\omega)k}.$$ (3.31)

The definition of collision temperature in Bird (1994) was based on the energy in all modes of the colliding molecules. However, as noted in the preceding section, the Larsen-Borgnakke energy redistribution is based on the sum of the translational energy and a single internal mode and the appropriate collision temperature is then

$$T_{coll} = \frac{\frac{1}{2}m_r c_r^2 + \varepsilon_{int}}{\left(\frac{5}{2} + \overline{\zeta_{int}} - \omega\right)k}. \tag{3.32}$$

The definition in Eqn. (3.31) is preferred because temperature is a measure of energy and, to define a temperature, the energy is divided by the product of the Boltzmann constant and the "effective" number of degrees of freedom that takes the degree of excitation into account. However, the Larsen-Borgnakke selection is based on the physical degrees of freedom, irrespective of the degree of excitation, and, for a partially excited mode, Eqn. (3.32) leads to a temperature that is too low.

After the introduction of the quantum vibration model, it was found that the vibrational temperature did not come to equilibrium with the other temperatures when the vibrational relaxation rate was based on the collision temperature rather than the macroscopic temperature. It was shown (Bird, 2002) that equipartition is achieved if the total collision energy is quantized in the same way as the energy of the relevant vibrational mode. It was then advocated that the collision temperature should be employed when setting all temperature dependent parameters. However, a subsequent study (Bird, 2009) showed that, while equilibrium is achieved with a quantized collision temperature, the vibrational relaxation is much faster than the nominal value.

The error in the relaxation rate occurs because the vibrational relaxation rate is proportional to a power of the temperature and, while the temperature T is equal to the collision temperature T_{coll}, T^B is not generally equal T_{coll}^B. The former is proportional to the mean value of c_r^{2B} in collisions, while the latter is proportional to the mean value of c_r^2 in collisions raised to the power B. From Eqn. (2.30)

$$\frac{\overline{c_r^{2B}}}{\left(\overline{c_r^2}\right)^B} = \frac{\Gamma\left(\frac{5}{2} - \omega + B\right)}{\left(\frac{5}{2} - \omega\right)^B \Gamma\left(\frac{5}{2} - \omega\right)}.$$

Therefore, if a relaxation collision number is AT^B, the number $A'T_{coll}^B$ leads to the correct value if

$$A'/A = \left(\frac{5}{2} - \omega\right)^B \Gamma\left(\frac{5}{2} - \omega\right)/\Gamma\left(\frac{5}{2} - \omega + B\right). \tag{3.33}$$

This means that, if the relaxation collision number is written as the product of a constant A and the temperature to the power B, Eqn. (3.33) suggests that it should be possible to alter the constant so that the collision temperature can be used in place of the macroscopic temperature. The ratio A'/A of the changed to the original constant is plotted in Figure 3.3 as a function of the power law B.

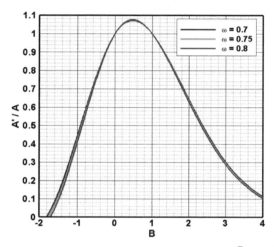

Fig. 3.3 The ratio A'/A that is requited for $A' T_{coll}^B = A T^B$.

The collision temperature may be employed when implementing a temperature dependent rotational relaxation rate because typical data (Lordi and Mates, 1970) for the variation of the rotational collision number with temperature suggests indices in the range 0 to 1. On the other hand, it is problematic to employ the collision temperature for vibration. The denominator of Eqn. (3.33) is infinity at $B = \omega - \frac{5}{2}$ and the correction ratio is then zero. The increase in vibrational relaxation collision number with decreasing temperature is such that the effective B for vibration is near the singularity.

In any case, the physical argument for the use of a collision temperature is in conflict with the earlier statement that a phenomenological model is not necessarily improved by attempts to increase the physical realism of the model. Therefore when using temperature dependent data in DSMC, it is now recommended that the macroscopic translational temperature be employed.

The Millikan-White (1963) theory predicts that the product of the pressure and the vibrational collision time is proportional to the exponential of a constant times the minus one third power of the temperature. This leads (Bird, 1994) to a vibrational collision number

$$Z_{vib} = \left(C_1/T^\omega\right)\exp\left(C_2 T^{-1/3}\right),$$

where C_1 and C_2 are constants. The values of the constants that best fit the experimental data were provided for typical gases, but a problem with Millikan-White data is that the collision number goes to an unphysical value less than unity before the dissociation temperature is reached. There is a strong case (see §3.6) for setting the vibrational collision number to unity at the characteristic temperature of dissociation Θ_d and it is then possible to specify the vibrational collision number through a single value $Z_{vib,ref}$ at the reference temperature T_{ref}. The expression for the vibrational collision number is then

$$Z_{vib} = \left(\frac{\Theta_d}{T}\right)^\omega \left\{Z_{vib,ref}\left(\frac{T_{ref}}{\Theta_d}\right)^\omega\right\}^{\left\{\left(\frac{\Theta_d}{T}\right)^{1/3}-1\right\}\bigg/\left\{\left(\frac{\Theta_d}{T_{ref}}\right)^{1/3}-1\right\}} . \tag{3.34}$$

3.6 Models for chemical reactions

Most simulations of chemical reactions in DSMC have employed the TCE method that was detailed in Bird (1994). This has now been superseded by the quantum-kinetic, or Q-K method and this will be implemented exclusively in the programs in this book. This theory (Bird, 2011) is based on the quantum vibration model that is presented in §3.4 and, because the programs employ the simple harmonic model for vibration, the presentation employs that model. However, just as the specialization of the quantum vibration model to the simple harmonic model just involved the replacement of ε_{vib} by $i k \Theta_{vib}$, the theory could be generalized by the reverse substitution.

Consider the serial application of the quantum L-B model to a vibrational mode of one of the molecules in a collision in which the sum of the relative translational energy and the pre-collision vibrational energy of the molecule and mode under consideration is E_c. The maximum vibrational level that can be selected is

$$i_{max} = \lfloor E_c / (k\Theta_{vib}) \rfloor. \tag{3.35}$$

Potential post-collision states i^* are chosen uniformly from the states equal to or below i_{max} and are selected through an acceptance-rejection routine with probability

$$P = (1 - i^* k\Theta_{vib} / E_c)^{3/2 - \omega}, \tag{3.36}$$

There is a continuous set (i.e. an infinite number) of levels beyond the dissociation limit so that, if i_{max} is above the dissociation level, it must be selected and dissociation occurs. The condition for **dissociation** of the molecule **AB** in the reaction **AB+T→A+B+T** is therefore

$$i_{max} > \Theta_d / \Theta_{vib}, \tag{3.37}$$

The vibrational collision number is set to unity at the dissociation limit so that the condition is applied in the DSMC collision routine prior to the Larsen-Borgnakke selection of a post-collision vibrational state and dissociation occurs whenever it is energetically possible. The Larsen-Borgnakke selection process is an even choice from the available states and the transition from a finite to an infinite number of states at the dissociation limit would lead to a discontinuity if the vibrational collision number was not unity at this limit.

The corresponding rate coefficient for dissociation in an equilibrium VHS gas is

$$k_f(T) = g R_{coll} \Upsilon(i_{max}) \tag{3.38}$$

The degeneracy g covers the case of a polyatomic molecule in which the reaction can occur through more than one mode with the same characteristic vibrational temperature. The collision rate parameter $R_{coll}^{AB,T}$ is the collision rate for collisions between gas species **AB** and **T** divided by the number density product. It can be written for an equilibrium VHS gas as

$$R_{coll} = \left(2\pi^{1/2} / \varepsilon\right)\left(r_{ref}^{AB} + r_{ref}^{T}\right)^2 \left(T / T_{ref}\right)^{1 - \omega^{AB,T}} \left(2kT_{ref} / m_r^{AB,T}\right)^{1/2}, \tag{3.39}$$

where r_{ref} is the molecular radius at temperature T_{ref}, m_r is the reduced mass and ε is a symmetry factor equal to 1 for unlike molecules and 2 for like molecules. The parameter $\Upsilon(i_{max})^{AB,T}$ is the fraction of collisions between **AB** and **T** that have sufficient energy to meet the condition in Eqn. (3.37). For an equilibrium VHS gas,

$$\Upsilon(i_{max}) = \frac{\sum_{i=0}^{i_{max}-1}\left\langle Q\left[\left(\tfrac{5}{2}-\omega^{AB,T}\right),\left\{(i_{max}-i)\Theta_{vib}^{AB}/T\right\}\right]\exp\left(-i\Theta_{vib}^{AB}/T\right)\right\rangle}{z_{vib}(T)^{AB}}. \tag{3.40}$$

$Q(a,x) = \Gamma(a,x)/\Gamma(a)$ is a form of the incomplete Gamma function and $z_{vib}(T) = 1/[1-\exp(-\Theta_{vib}/T)]$ is the contribution of the relevant mode to the vibrational partition function for the simple harmonic oscillator model. While this rate equation is not required for the implementation of the Q-K model in DSMC programs, it allows comparisons with the rate equations based on experiment.

Because Eqn. (3.40) involves a summation over the pre-dissociation vibrational levels and the collision rate parameter is independent of this level, it is possible to write an expression for the vibrational level-resolved dissociation rate. The ratio of the level j rate to the overall rate is

$$\frac{k_f^j(T)}{k_f(T)} = \frac{Q\left[\left(\tfrac{5}{2}-\omega^{AB,T}\right),\left\{(i_{max}-j)\Theta_{vib}^{AB}/T\right\}\right]\exp\left(-j\Theta_{vib}^{AB}/T\right)}{\sum_{i=0}^{i_{max}-1}\left\langle Q\left[\left(\tfrac{5}{2}-\omega^{AB,T}\right),\left\{(i_{max}-i)\Theta_{vib}^{AB}/T\right\}\right]\exp\left(-i\Theta_{vib}^{AB}/T\right)\right\rangle}. \tag{3.41}$$

The DSMC condition for **recombination** in a collision between species **A** and **B** is that there is another atom or molecule within the "ternary collision volume" V_{coll}. The probability of recombination is therefore

$$P_{rec} = nV_{coll} \tag{3.42}$$

where n is the number density. The corresponding rate equation is

$$k_r(T) = R_{coll}^{A,B}\, nV_{coll}. \tag{3.43}$$

The ratio $k_f(T)/k_r(T)$ from Eqns. (3.38) and (3.43) may be equated to the equilibrium constant from statistical mechanics in order to determine the ternary collision volume. It has been found that a volume in the form

$$V_{coll} = a\left(T/\Theta_{vib}\right)^b V_{ref} \qquad (3.44)$$

allows an almost exact fit of this ratio to the equilibrium constant. The reference volume V_{ref} is most conveniently set to the volume of the sphere with radius equal to the sum of the reference radii of the three molecules.

Now consider **exchange and chain reactions**. An exchange reaction has one stable molecule and one radical both before and after the reaction. A reaction between two stable molecules that leads to one or more radicals is, like thermal dissociation, a chain-initiating reaction. When a pre-reaction pair comprised of one stable molecule and one radical leads to two radicals, it is a chain-branching reaction, while the opposite is a chain-terminating reaction.

Consider the forward and reverse reaction pair **A+B** $\leftrightarrow$ **C+D** in which **A** and **C** are the molecules that split, while **B** and **D** are either molecules or atoms. The phenomenological DSMC procedure for these reactions is analogous to that for the dissociation model. For the forward or endothermic reactions, the reaction probability is equal to that of Eqn. (3.36) with i^* set to the i_{max} defined by Eqn. (3.35) and with E_c replaced by the activation energy that applies to the reaction. This is a relative probability that must be normalized through division by the sum of probabilities from the ground state to i_{max}. Therefore

$$P_{reac} = \left(1 - i_{max}k\Theta_{vib}/E_c\right)^{3/2-\omega} / \sum_{i=0}^{i_{max}}\left(1 - ik\Theta_{vib}/E_c\right)^{3/2-\omega}.$$

This is satisfactory only if the level i_{max} is very large in comparison with unity. Otherwise, the use of discrete levels leads to unacceptably abrupt changes in the rate coefficients and it is impossible obtain a near exact agreement between the ratio of forward to reverse rate coefficients and the equilibrium constant. The level i_{max} is the highest level with energy below the activation energy E_a and a smooth curve is obtained for the rate coefficients if the energy $i_{max}k\Theta_{vib}$ in the numerator is replaced by E_a. The DSMC probability of an exchange or chain reaction in a collision with $E_c > E_a$ is then

$$P_{reac} = \left(1 - E_a/E_c\right)^{3/2-\omega} / \sum_{i=0}^{i_{max}}\left(1 - ik\Theta_{vib}/E_c\right)^{3/2-\omega}. \qquad (3.45)$$

The summation in the denominator is unity when E_a/k is less than Θ_{vib} and only a few terms need be calculated when the activation energy is small. While there are many terms when the activation energy is large, the sum is then a near linear function of $E_c - E_a$ and the summation may be replaced by a simple expression.

The selection of possible post-collision vibrational levels through Eqn. (3.45) is from an equilibrium distribution so that the rate coefficient is simply the product of the degeneracy g, the collision rate parameter and the fraction of the vibrational level i_{max} in the Boltzmann distribution. The term $i_{max}k\Theta_v$ in the exponential should be replaced by E_a in order to maintain consistency with Eqn. (3.45) and the normalizing factor is simply the vibrational partition function so that

$$k_f(T) = g^A \, R_{coll}^{A,B} \exp\left(-E_a^{A,B} / (kT)\right) / z_{vib}(T)^A. \qquad (3.46)$$

Bird (2011) presented an exact kinetic theory result for this rate coefficient and it was shown to lead to results that are identical to those from Eqn. (3.46). The kinetic theory result involves a summation over the vibrational states and the two expressions may be combined to obtain an equation for the level-resolved rates. This leads to an unwieldy result and it is desirable to simplify it through the introduction of dimensionless parameters that are distinguished from the corresponding dimensioned parameters by a prime. These are the relative translational energy in the collision $E_{tr}' = m_r c_r^2 / (2kT)$, the characteristic vibrational temperature $\Theta_{vib}' = \Theta_{vib} / T$, the energy of the vibrational mode that is broken in the reaction $j\Theta_{vib}'$, the activation energy of the reaction $E_a' = E_a / (kT)$, the highest vibrational level with energy below the sum of the relative translational energy and the energy in the vibrational that is broken in the reaction $i_{max} = \lfloor E_{tr}' / \Theta_{vib}' + j \rfloor$ and, finally, the energy that must be added to the vibrational energy $j\Theta_v'$ in order to reach the activation energy $E_r' = E_a' - j\Theta_v'$ if positive, otherwise 0. The Q-K result for the ratio of the rate coefficient for molecules with vibrational level j to the overall rate coefficient for the reaction is then

$$\frac{k_f^j(T)}{k_f(T)} = \frac{z_v(T)}{\Gamma(5/2-\omega)\exp\left(E_a'\right)} \times$$

$$\times \int_{E_r'}^{\infty} \frac{\left[E_{tr}'\left\{1 - E_a' / \left(E_{tr}' + j\Theta_{vib}'\right)\right\}\right]^{3/2-\omega}}{\sum_{i=0}^{i_{max}}\left[1 - i\Theta_{vib}' / \left(E_{tr}' + j\Theta_{vib}'\right)\right]^{3/2-\omega}} \exp\left(-E_{tr}'\right) dE_{tr} \quad , \quad (3.47)$$

where $z_{vib}(T) = 1/\left[1 - \exp\left(-\Theta_{vib}'\right)\right]$ is the vibrational partition function.

The summation is related to the Hurwitz zeta function and, if the properties of this function were as well-known as those for the Gamma functions, Eqn. (3.47) could almost certainly be simplified.

The rate coefficient of Eqn. (3.46) may be combined with the equilibrium constant to obtain the following expression for the ratio of the reverse to the forward reaction rate coefficient

$$\frac{k_r(T)}{k_f(T)} = \left(\frac{m^A m^B}{m^C m^D}\right)^{3/2} \frac{{}^A z_{rot} {}^B z_{rot} {}^A z_{vib} {}^B z_{vib} {}^A z_{el} {}^B z_{el}}{{}^C z_{rot} {}^D z_{rot} {}^C z_{vib} {}^D z_{vib} {}^C z_{el} {}^D z_{el}} \exp\left(\frac{E_a}{kT}\right). \quad (3.48)$$

The activation energies for the endothermic exchange and chain reactions may be higher, by the energy barrier E_b, than the increase in the heats of formation. Similarly, there may be an energy barrier to an exothermic reverse reaction. This is taken into account when Eqn. (3.48) is used to determine the a' and b' coefficients in the following equation for the reverse reaction probability

$$P_{reac} \equiv k_r(T)/R_{coll}^{C,D} = a' T^{b'} \exp\left[-E_b/(kT)\right]. \quad (3.49)$$

The Larsen-Borgnakke distributions that satisfy detailed balance in a non-reacting gas do not lead to detailed balance when there are reactions with finite activation energies. Detailed balance is satisfied only if the distribution for each post-reaction energy mode of all species matches the corresponding pre-collision distributions of the depleting reactions. Equation (3.41) may be used to set the post-reaction distribution of vibrational energy of the molecule that is formed in a recombination. Similarly, Eqn. (3.47) may be used to set the post-reaction vibrational distributions in reverse exchange and chain reactions. Because the energy barrier of Eqn. (3.49) need not be taken into account in the selection of the collision pair for an exothermic reaction, the L-B distributions continue to be valid for the post-reaction vibrational states of the forward reaction products.

References

Bergemann, F. (1994). Gaskinetische Simulation von kontinuumsnahen ... von Wandkatalyse, Ph.D. Thesis, Institut für Strömungsmechanik, Göttingen, Germany.

Bird, G. A. (1970). Aspects of the Structure of Strong Shock Waves. *Phys. Fluids* **11**, 1173-1177.

Bird, G. A. (1981). Monte Carlo simulation in an engineering context. *Progr. Astro. Aero.* **74**, 239.

Bird, G. A. (1994). *Molecular Gas Dynamics and the Direct Simulation of Gas Flows,* Clarendon Press, Oxford.

Bird, G. A. (2002). A criterion for the breakdown of vibrational equilibrium in expansions. *Phys. Fluids* **14**, 1732.

Bird, G. A. (2009). A Comparison of Collision Energy-based and Temperature-based Procedures in DSMC. In *Rarefied Gas Dynamics* (Ed. T. Abe), 245-250, American Institute of Physics, Melville, New York.

Bird, G. A. (2011). The Q-K model for gas phase chemical reaction rates. *Phys. Fluids* **23**, 106101.

Bird, G. A. (2012). Setting the post-reaction internal energies in DSMC chemistry simulations. *Phys. Fluids* **24,** 127104.

Borgnakke, C. and Larsen, P. S. (1974). Statistical collision model for simulating polyatomic gas with restricted energy exchange. In *Rarefied Gas Dynamics* (ed. M. Becker and M. Fiebig), **1**, Paper A7, DFVLR Press, Porz-Wahn, Germany.

Chapman, S. and Cowling, T. G. (1970). *The Mathematical Theory of Non-uniform Gases* 3rd edn., Cambridge University Press.

Gallis, M. A., Bond, R. B. and Torczynski, J. R. (2010). A kinetic theory approach for computing chemical-reaction rates in upper-atmosphere hypersonic flows, *J. Chem. Phys.* **131**, 124311.

Levine, R. D. and Bernstein, R. B. (1974). "Energy disposal and energy consumption in elementary chemical reactions. Information theoretic approach," Acc. Chem. Res. **7**, 393.

Levine, R. D. and Bernstein, R. B. (2005). *Molecular Reaction Dynamics and Chemical Reactivity*, Cambridge University Press

Lordi, J. A. and Mates R. E. (1970) Rotational Relaxation in nonpolar diatomic gases, *Phys Fluids* **13**, 291-308.

Millikan, R. C. and White, D. R. (1963) Systematics for vibrational relaxation, *J Chem. Phys.* **39**, 3209-3213.

Pullin, D. I. (1979). Kinetic models for polyatomic molecules with phenomenological energy exchange, *Phys. Fluids* **21**, 209-216.

Sharipov, F. and Strapasson, J. L. (2012). Direct simulation Monte Carlo for an arbitrary intermolecular potential, *Phys. Fluids* **24**, 011703.

4

DSMC PROCEDURES

4.1 Generation of reference states

Almost all DSMC programs involve the generation of a uniform equilibrium gas and most involve a boundary at which a uniform gas stream enters the flow. With the exception of flows at high density and with very small linear dimensions, each simulated molecule represents an extremely large number of real molecules. With this in mind, the initial and continuing "randomness" of the flow should be reduced wherever possible. For example, the molecules that are set at zero time should be uniformly spaced rather than each molecule being placed at a random location within the flowfield. Similarly, molecules that enter across a segment of boundary should be evenly spaced along the segment.

Many applications, including all studies of homogeneous gases, involve a fixed number of molecules with assumed zero stream velocity at a prescribed number density and pressure. That is, the momentum summed over all molecules should be zero in each direction and the sum of the molecular energies should be that corresponding to the specified temperature. The number density will be the desired value but the momentum in each direction and the overall energy is subject to a random walk. Each molecule that is added represents a step in the random walk and the mean deviations from the nominal values increase as the square root of the number of molecules. To avoid these random walk discrepancies, the procedures keep track of the overall momentum in each direction and alternative pairs of random velocity components are generated. Should the total momentum in one direction be positive, the smaller value in the pair for that direction is chosen, and vice versa. In order to minimize the energy deviation as well as the momentum deviations, alternative momentum pairs have to be generated and the corresponding choices made with regard to the total energy. Alternative internal energies may be included in the low-discrepancy procedure for the overall temperature.

The procedures for the sampling of typical values from a prescribed distribution function have been discussed in Appendix C of Bird (1994). The distribution of the variate x is such that the probability of a value between x and $x + dx$ is $f_x dx$ and, if the range of x is from a to b, the distribution is normalized such that its integration from a to b is unity. The definition of the **cumulative distribution function** is

$$F_x = \int_a^x f_x \, dx \,. \tag{4.1}$$

This function may be equated to a random fraction R_F in order to find the value of the cumulative distribution that corresponds to a representative value of the variate x. Therefore, if the expression for F_x as a function of x can be inverted, it provides an expression for a representative value of x as a function of R_F. For example, it was employed in Chapter 3 to obtain Eqn. (3.26) for the post-collision energy in an internal mode with two degrees of freedom.

The initial velocity components of the simulated molecule in an equilibrium gas are generated from the distribution function of Eqn. (2.9). The cumulative distribution function is

$$F_{\beta u'} = \left\{ 1 + \mathrm{erf}\left(\beta u' \right) \right\} / 2$$

and cannot be inverted.

The most common alternative to the inverse cumulative distribution procedure is the **acceptance-rejection** procedure. The ratio of the probability of a particular value of the variate to the maximum probability is equal to the ratio of the distribution function to the maximum value of the distribution. *i.e.*

$$P/P_{max} = f_x / f_{x,max} \,. \tag{4.2}$$

A random fraction is generated and a value of the variate x is chosen on the assumption that it is uniformly distributed between a and b. *i.e.*

$$x = a + R_F \left(b - a \right). \tag{4.3}$$

The value of P / P_{max} for this value of x is calculated and is compared with a new random fraction R_F. This value of x is accepted if $P / P_{max} > R_F$ and rejected if $P/P_{max} < R_F$. If it is rejected, a new value of x is generated from Eqn. (4.3) and the process is repeated until a value is accepted.

A problem with the application of the acceptance-rejection procedure to the velocity components is that they range from minus infinity to infinity. Cut-off limits have to be applied to the allowed range of values for $\beta u'$, $\beta v'$, and $\beta w'$ and, the wider the limits, the less efficient the procedure. Figure 2.1 shows that limits of ±4 would be required to keep the errors that are inevitably produced by cut-offs down to about one part in a million. Fortunately, it is possible to apply the inverse cumulative distribution procedure to the generation of pairs of representative velocity components. A pair of independent random fractions is employed to set radial coordinates in two-dimensional velocity space. *i.e.*

$$r = \left\{-\ln(R_F)\right\}^{1/2}/\beta \quad \text{and} \quad \theta = 2\pi R_F . \tag{4.4}$$

The two representative **translational velocity components** are then

$$v' = r\cos\theta \quad \text{and} \quad w' = r\sin\theta . \tag{4.5}$$

The generation of pairs of velocity components is particularly useful when the low-discrepancy procedure is employed in the generation of the initial state in order to counter the random walks in the total momentum and energy.

For the generation of representative initial values of the rotational energy, the implied constant in the distribution function of Eqn. (2.11) must be evaluated from the normalization condition that its integration from 0 to ∞ must be unity. The result is

$$f_{\varepsilon_{rot}/(kT)} = \left\{\varepsilon_{rot}/(kT)\right\}^{\zeta/2-1} \exp\left\{-\varepsilon_{rot}/(kT)\right\}/\Gamma(\zeta/2) \tag{4.6}$$

and the cumulative distribution function for the special case of two rotational degrees of freedom is

$$F_{\varepsilon_{rot}/(kT)} = 1 - \exp\left\{-\varepsilon_{rot}/(kT)\right\} . \tag{4.7}$$

Noting that $R_F \equiv 1 - R_F$, this equation shows that a representative value of the **rotational energy of a diatomic molecule** in an equilibrium gas is

$$\varepsilon_{rot} = -\ln(R_F)kT . \tag{4.8}$$

An acceptance-rejection procedure is required for other than a diatomic gas and the ratio of the probability to the maximum probability for the **rotational energy of a polyatomic molecule** is

$$\frac{P}{P_{max}} = \left\{ \frac{\varepsilon_{rot}}{(\zeta_{rot}/2 - 1)kT} \right\}^{\zeta_{rot}/2 - 1} \exp\left(\frac{\zeta_{rot}}{2} - 1 - \frac{\varepsilon_{rot}}{kT} \right). \qquad (4.9)$$

The simple harmonic model will be employed for the vibrational mode and a level will be stored for each vibrational mode of a simulated molecule with one or more vibrational modes. Each vibrational mode contributes two degrees of freedom and Eqn. (4.9) for two degrees of freedom is consistent with the Boltzmann distribution. The simplest procedure for generating the initial distribution of levels is to apply Eqn. (4.8) for the vibrational energy and to truncate this energy to the energy $ik\Theta_{vib}$ of the highest available level. A **representative vibrational level** of a mode with a characteristic vibrational temperature Θ_{vib} is

$$i = \lfloor -\ln(R_F) T/\Theta_{vib} \rfloor. \qquad (4.10)$$

The electronic levels are irregularly spaced and few in number. A very large fraction of the molecules are initially in the ground state and a direct sampling from the Boltzmann distribution is feasible. For a given temperature, the fraction of molecules in each level is calculated from Eqn. (2.12). A random fraction is generated and the molecule is initially placed in the ground state. If the random fraction is below the fraction in the ground state, it remains in the ground state. If not, the level is progressively increased until the cumulative fraction exceeds the random fraction. The **representative electronic level** j is therefore obtained when

$$\sum_{i=0}^{j} N_i/N > R_F . \qquad (4.11)$$

For a gas mixture, the preceding equations are applied separately to each molecular species in order to generate the initial equilibrium gas. The number of simulated molecules that is generated per unit volume is simply the number density divided by the number F_N of real molecules that are represented by each simulated molecule.

Should the boundary conditions of an application involve an interface with an equilibrium gas, is also necessary to generate the molecules that move into the flowfield across the interface. The special case of an interface with a stationary gas is particularly important because the same equations apply to the molecules that are diffusely reflected from a solid surface.

Consider a boundary interface with a stationary gas of number density n and temperature T. The number flux of molecules of mass m that cross the interface per unit area per unit time is given by Eqn. (2.16) as

$$\dot{N} = n/(2\pi^{1/2}\beta) \quad \text{where} \quad \beta = \{m/(2kT)\}^{1/2}. \tag{4.12}$$

The number of simulated molecules is again obtained by dividing by F_N. In the case of the diffusely reflected molecules, T becomes the surface temperature and n is an equivalent number density that is determined by the inward number flux to the surface. For generating the velocity components of the molecules crossing the boundary, Cartesian coordinates are defined as in §2.2 with the u component normal to the boundary or surface element and the v and w components parallel to the element. The parallel components have no influence on the flux through the element and can therefore be generated by Eqns. (4.4) and (4.5). The probability of a molecule crossing the element is proportional to u and the distribution function for this component is

$$f_u \propto u \exp(-\beta^2 u^2).$$

This must be normalized such that its integration from 0 to ∞ is unity and may be transformed to a distribution for $\beta^2 u^2$. The result is

$$f_{\beta^2 u^2} = \exp(-\beta^2 u^2). \tag{4.13}$$

This distribution for $\beta^2 u^2$ is identical to that for $\varepsilon_{rot}/(kT)$ in Eqn. (4.6) when $\zeta = 2$. The inverse cumulative procedure then gives the representative value of the **normal velocity component in a stationary gas** as

$$u = \{-\ln(R_F)\}^{1/2}/\beta. \tag{4.14}$$

Should there be a component u_0 of the stream velocity normal to the element, the number flux across the element is, from Eqn. (2.15)

$$\dot{N} = n\left[\exp(-s_n^2) + \pi^{1/2}s_n\{1 + \operatorname{erf}(s_n)\}\right]/(2\pi^{1/2}\beta),$$

where   (4.15)

$$s_n = u_0\beta.$$

The thermal velocity component $u' = u - u_0$ and the distribution function for $\beta u'$ is

$$f_{\beta u'} = (\beta u' + s_n) \exp(-\beta^2 u'^2).$$

The acceptance-rejection procedure is required for the generation of representative velocity components from this distribution. The probability ratio is

$$\frac{P}{P_{max}} = \frac{2(\beta u' + s_n)}{s_n + (s_n^2 + 2)^{1/2}} \exp\left[\frac{1}{2} + \frac{s_n}{2}\left\{s_n - (s_n^2 + 2)^{1/2}\right\} - \beta^2 u'^2\right] \quad (4.16)$$

and a cut-off of some multiple of $\beta u'$ must be applied on either side of s_n. This procedure has been employed in most DSMC programs, but Garcia and Wagner (2006) have presented an exact method that does not require cut-offs. The exact procedure is considerably more complex and a calculation has been made with the **DS1**[†] program to determine whether it is necessary to replace the simpler procedure.

Consider a one-dimensional plane flow of a hard sphere gas at a number density of 1×10^{23} m⁻³, temperature of 300 K, and a velocity of 500 m/s. The flow has an extent of 20 mean free paths and both the upstream and downstream boundaries are stream interfaces with the number of entering molecules calculated from Eqn. (4.15). The normal velocity components of the entering molecules are calculated by the acceptance-rejection procedure based on Eqn. (4.16) and the two parallel components are generated from Eqns. (4.4) and (4.5).

Table 4.1 The errors that are introduced by velocity cut-offs.

Cut-off $(\beta u')$	n (m⁻³)	T (K)	u_0 (m/s)
± 2	1.007×10^{23}	296.0	496.5
± 2.5	1.0067×10^{23}	299.40	499.62
± 3	1.0000×10^{23}	299.96	500.00
± 3.5	0.9999×10^{23}	299.99	500.03

[†] The DS1 program is the basis of the DSMC program that forms an integral part of this book. It has been available since 2010 as freely downloadable source and executable code from www.gab.com.au. Version 1.14 was employed for the calculations that are reported in Table 4.1.

The effect of the velocity cut-offs on the number density, temperature and velocity are shown in Table 4.1. The results indicate that the cut-off of $\pm 3\beta u'$ that has traditionally been used is adequate, but the **DSMC.F90** program employs $\pm 4\beta u'$. Figure 2.1 shows that only one entry molecule in 10^7 falls outside these limits. This higher limit was not tested because, even though the sample size was in excess of 10^{12}, random walk effects prevent a better than four to five significant figure accuracy in the calculations.

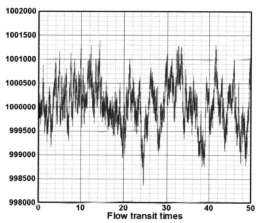

Fig. 4.1 A typical history of the total number of simulated molecules.

The calculations for Table 4.1 employed a nominal one million simulated molecules but, as shown for a typical case in Fig. 4.1, the actual number fluctuated by approximately $\pm 0.15\%$. This is because, while the inflow molecules at each boundary are uniformly distributed in time, the outflow molecule fluxes are subject to scatter and there is a random walk in the total number of simulated molecules. The characteristics of these random walk fluctuations differ from the Poisson distribution for the number of molecules in a gas element that contains an average of a million molecules. The average magnitude of the fluctuations is lower in that only 68% of a Poisson sample would be between 999,000 and 101,000. However, the convergence of the results with increasing sample size is adversely affected by the increasing average time interval between "zero-crossings" in the random walk.

4.2 Collision procedures

The flowfield is divided into collision cells and the molecules in a cell are regarded as being representative of the molecules at the location of the cell and all combinations of two molecules have traditionally been regarded as potential binary collision pairs. A time parameter t_{cell} is assigned to each collision cell and collisions appropriate to Δt_{cell} are calculated whenever t_{cell} falls $\frac{1}{2}\Delta t_{cell}$ behind the overall flow time. The time step Δt_{cell} is set to a specified small fraction of the sampled mean collision time in the sampling cell in which the collision cell lies. A time parameter is associated with every molecule and a similar procedure is applied to the molecule moves. A move time step equal to the collision time step would automatically satisfy the condition that the move distance should be small in comparison with the local mean free path. However, there are often practical considerations, especially in high Knudsen number flows, that call for a move time step that is much smaller than the collision time step. In addition, the overall time step should not be larger than the smallest collision time step.

The probability of a collision is proportional to the product of the relative speed of the molecules and the collision cross-section. Random pairs of molecules are selected and the collision pairs are chosen from these through an acceptance-rejection procedure based on the probability ratio

$$\frac{P}{P_{max}} = \frac{\sigma c_r}{(\sigma c_r)_{max}}. \tag{4.17}$$

The maximum value of the product of the total collision cross-section and the relative speed is initially set to a reasonably large value and, should a larger value be encountered during a calculation, it is increased

The determination of the number of possible collision pairs to be considered in a collision cell over the time step for that cell is the critical step in the overall procedure. The NTC scheme has been shown (Bird, 1994) to lead to the correct collision rate. This states that the number of possible collision pairs to be selected in a cell with N simulated molecules is

$$\frac{1}{2} N \overline{N} F_N (\sigma c_r)_{max} \Delta t_{cell} / V_{cell},$$

where V_{cell} is the volume of the collision cell.

This expression contains the mean value of the number of molecules. This takes some time to establish and may change rapidly in an unsteady flow. Also, it has a finite value when there is only one molecule in a collision cell and this number has to be carried forward along with the remainder when the number of selections is truncated to an integer. While these are minor problems, they may be avoided if advantage is taken of a mathematical property of the Poisson distribution that describes the fluctuations in the molecule number. This is that

$$\overline{N(N-1)} \equiv \overline{N}\overline{\overline{N}}$$

and the collision rate is unchanged if the number of pair selections is

$$\tfrac{1}{2} N(N-1) F_N (\sigma c_r)_{max} \Delta t_{cell} / V_{cell}. \tag{4.18}$$

Note that the parameter $(\sigma c_r)_{max}$ appears in the denominator of Eqn. (4.17) and the numerator of Eqn. (4.18) so that the collision rate is not affected by its actual value. The parameter $(\sigma c_r)_{max}$ may increase slightly during a run when a collision occurs with an exceptionally high relative speed. Because a VHS cross-section that is inversely proportional to a fractional power of the relative speed appears always in a product with the relative speed, the singularity at zero relative speed is of no consequence.

Early DSMC programs employed a single set of cells and the number of simulated molecules per cell was generally around twenty or thirty. It soon became clear that the mean spacing between collision pairs was too large in comparison with the mean free path and this led to the introduction of sub-cells. These have now been superseded by separate sets of collision and sampling cells and the emphasis has been on the reduction of the mean separation between the molecules that are accepted as collision pairs. The early DSMC programs regarded the molecules within a cell as representative of the molecules at the location of the cell and the relative locations of the molecules within the cell were ignored. This continues to be an option with regard to the selection of collision pairs from collision cells. The smaller the size of the collision cells, the smaller the ratio of the mean separation of the collision pairs to the local mean free path. The benefits from a reduction in this ratio are so large that there is now an option to select the collision pairs from the nearest-neighbour pairs within the cell.

The implementation of Eqn. (4.18) requires the determination of the collision cross-section. All the DSMC implementations in the programs associated with this book employ either the VHS or VSS models that reproduce the measured coefficients of viscosity and heat conduction and, in the case of the VSS model, also the self-diffusion coefficient. In both cases, the molecular diameter is determined, through the Chapman-Enskog theory, from the measured values of the coefficient of viscosity. The relevant equation for the VSS model is (3.19) and, for $\alpha = 1$, this reduces to Eqn. (2.43) for the VHS model. The cross-section σ is related to the molecular diameter by $\sigma = \pi d^2$ and the diameter at a particular value of relative speed c_r is given by Eqn. (2.32). A problem is that Eqns. (2.43) and (3.19) are just the first term in the Chapman-Enskog solution based on Sonine polynomials. The higher order corrections are of the order of one or two percent for viscosity and heat conduction, but are up to five percent for the diffusion coefficient. The corrections have been discussed by Chapman and Cowling (1970) for hard sphere and power-law models and may be estimated for the VHS and VSS models.

The molecular data that is required for the implementation of Eqn. (4.18) in a simple VHS or VSS gas is the molecular mass, the molecular diameter at a specified reference temperature, the temperature power law to which the viscosity coefficient is proportional, and the VSS deflection parameter α that is unity in a VHS gas. In the case of gas mixtures, the programs associated with this book employ the average molecular properties for the calculation of the cross-sections of collisions between unlike molecules. Bird (1994, §12.3) made calculations, using the VSS model, for the viscosity of a 6:4 helium-argon mixture for which the measured viscosity coefficient is 4% higher than the pure argon coefficient, even though the pure helium coefficient is 14% less than that of argon. Both the calculations that employed the mean values and similar calculations that directly employed the measured viscosity coefficient in the mixture produced similar results.

A difficulty in applications that involve large temperature differences is that measured values of the viscosity-temperature index ω are available only at temperatures of the order of standard temperature and predicted cross-sections at very high temperatures are unlikely to be accurate. Isolated high temperature data may be available and could be matched through a temperature dependent ω.

4.3 Molecule move procedures

A time parameter t_{mol} is associated with each simulated molecule and the molecule is moved through a distance appropriate to Δt_{mol} whenever t_{mol} falls $\frac{1}{2}\Delta t_{mol}$ behind the overall flow time. This is consistent with the collision cell times and time steps when Δt_{mol} is set to a small fraction of the local mean collision time so that the distance moved is similarly related to the local mean free path. This is one of the options that can be chosen by setting the computational parameter **IMTS** to 1. However, in the case of high Knudsen number flows, this can lead to molecule move distances that are excessively long with regard to the detection of the intersection of molecule trajectories with surfaces and flow boundaries. The parameter **IMTS** has therefore been assigned the default value of 0 and this causes the move time step to be equal to the overall times. The overall time step is the minimum value, over all collision cells, of the collision time step. The overall time step has the same value over the flowfield and is consistent with the fixed time step that has traditionally been employed in DSMC. The variable time step option should be regarded as a "developmental option" that is subject to continuing refinement.

Most calculations are made in a laboratory frame of reference that is free from external force fields and the molecules move in straight lines between collisions. In the absence of a boundary or surface interaction, the change in the position vector of the molecule is therefore

$$\Delta \mathbf{s} = \mathbf{c} \Delta t_{mol} . \tag{4.19}$$

The molecule is initially located at x_i, y_i, z_i, $\Delta \mathbf{s}$ has components Δx, $\Delta y, \Delta z$, and the direction cosines of the trajectory are $l = \cos^{-1}(\Delta x / s)$, $m = \cos^{-1}(\Delta y / s)$, $n = \cos^{-1}(\Delta z / s)$. A particularly useful theorem is that the distance from the initial location to the intersection of the line

$$x = x_i + l s$$

$$y = y_i + m s \tag{4.20}$$

$$z = z_i + n s$$

with the quadric surface

$$f(x,y,z) \equiv a_{11}x^2 + a_{22}y^2 + a_{33}z^2 + 2a_{23}yz + 2a_{31}zx + 2a_{12}xy$$
$$+ 2a_{14}x + 2a_{24}y + 2a_{34}z + a_{44} = 0 \qquad (4.21)$$

is given by the roots of the quadratic

$$A_1 s^2 + 2A_2 s + A_3 = 0, \qquad (4.22)$$

where

$$A_1 = a_{11}l^2 + a_{22}m^2 + a_{33}n^2 + 2a_{23}mn + 2a_{31}nl + 2a_{12}lm,$$

$$A_2 = l(a_{11}x_i + a_{12}y_i + a_{13}z_i + a_{14}) + m(a_{21}x_i + a_{22}y_i + a_{23}z_i + a_{24})$$
$$+ n(a_{31}x_i + a_{32}y_i + a_{33}z_i + a_{34})$$

and

$$A_3 = f(x_i, y_i, z_i).$$

If any real positive root s of Eqn. (4.22) is less than $|\Delta s|$, the molecule collides with the surface and the root can be substituted into Eqn. (4.20) to determine the coordinates of the point of collision. Should there be two positive real roots, the smaller one is relevant.

Note that three-dimensional quadric surfaces include planes so that the theorem can be applied to the individual triangles of general triangulated three-dimensional shapes. There will generally be possible intersection with more than one triangle in three dimensions or polyline segments in two dimensions. There are simple tests to determine whether intersection points are within triangles or line segments and the collision point is the nearest valid intersection. In addition, Eqn. (4.21) can be applied to conics and straight lines in two dimensions and even to points in one dimension. However, trivial cases such the interaction of molecules with bounding planes of symmetry are generally dealt with from first principles. For example, consider a plane of symmetry normal to the x axis and located at x_b. A plane of symmetry is equivalent to a specularly reflecting surface so that the u component of velocity is reversed and the other components are unchanged. If a molecule initially at x_i crosses the wall, its location after reflection is

$$x = 2x_b - x_i . \qquad (4.23)$$

Solid surfaces are generally specified as diffusely reflecting and the properties of the reflected molecules are as if they were effusing from a fictitious gas at the surface temperature on the opposite side of the surface. This equivalence was noted in §4.1 and the normal component of velocity of the reflected molecule is generated from Eqn. (4.14). The two parallel components are those of a stationary equilibrium gas and are generated from Eqns. (4.4) and (4.5).

Advantage is taken of flow symmetries when calculating cylindrical, spherical and axially-symmetric flows. An axially-symmetric flow has the axis in the x direction and the y axis is in the radial direction. The calculation is in the zero azimuth angle plane and, because the molecules move in three dimensions, the molecules move off this plane. The new radius becomes the new y coordinate and the v and w velocity components must be transformed as the new radius is moved back to the zero azimuth plane. For the one-dimensional cylindrical and spherical flows, the x axis is in the radial direction. Again, the molecules move off the axis and the new radius is becomes the new x coordinate. As with axially-symmetric flows, a transformation must be applied to the velocity components to allow for the positional change.

First consider cylindrical flow with the axis of the cylinder lying along the y axis. A molecule at x_i with velocity components u_i, v_i, w_i moves, in a time step, through $\Delta x, \Delta y, \Delta z$. The new radius is then

$$x = \sqrt{(x_i + \Delta x)^2 + \Delta z^2} \; . \tag{4.24}$$

The transformed values of the radial and circumferential velocity components are

$$u = \left\{ u_i (x_i + \Delta x) + w_i \Delta z \right\} / x$$

and
$$\tag{4.25}$$

$$w = \left\{ w_i (x_i + \Delta x) - u_i \Delta z \right\} / x \; .$$

The new radius in a spherical flow is

$$x = \sqrt{(x_i + \Delta x)^2 + \Delta y^2 + \Delta z^2} \; , \tag{4.26}$$

and the new radial velocity component is

$$u = \left\{ u_i (x_i + \Delta x) + \sqrt{v_i^2 + w_i^2} \sqrt{\Delta y^2 + \Delta z^2} \right\} / x \; . \tag{4.27}$$

The velocity component normal to the radial velocity is

$$u_n = \left\{ \sqrt{v_i^2 + w_i^2} \left(x_i + \Delta x \right) - u_i \sqrt{\Delta y^2 + \Delta z^2} \right\} / x .$$

The direction of this velocity is uniformly distributed in a spherical flow, so an angle ϕ is chosen at random between 0 and 2π and

$$v = u_n \sin\phi, \qquad w = u_n \cos\phi. \tag{4.28}$$

The axially-symmetric case is similar to the cylindrical flow, but with axes in different directions, and Eqn. (4.24) becomes

$$y = \sqrt{\left(y_i + \Delta y \right)^2 + \Delta z^2} . \tag{4.29}$$

Eqn. (4.25) for the transformed values of the radial and circumferential velocity components becomes

$$u = \left\{ u_i \left(y_i + \Delta y \right) + w_i \Delta z \right\} / y$$

and   (4.30)

$$w = \left\{ w_i \left(y_i + \Delta y \right) - v_i \Delta z \right\} / y .$$

Should there be an external force, the molecular paths between collisions are parabolic and there is a change in the molecular speed. For example, an acceleration a in the x direction can be handled through the elementary relations

$$\Delta x = u \Delta t_{mol} + \tfrac{1}{2} a \Delta t_{mol}^2$$

and   (4.31)

$$\Delta u = a \Delta t_{mol} .$$

Examples of the application of DSMC to gravitational flows have been provided by Bird (1994, §12.13).

Calculations for rotating flows can be made in either the laboratory frame of reference in which the molecules move in straight lines or in a rotating frame of reference in which the molecules are subject to centrifugal and Coriolis accelerations. Bird (1994) §12.14 presents a DSMC simulation of a gas centrifuge that is calculated in the laboratory frame of reference. The outer surface rotates and the molecule trajectories are straight. A similar procedure has been adopted in the many studies of centrifugal instabilities in Taylor-Couette flows. It is not always possible to adopt this approach when there is a combination of rotating and stationary surfaces as, for example, in a turbomolecular pump. The essentially unsteady nature of DSMC permits a calculation with all blades regarded as stationary. This avoids dynamic grids and, instead of moving rotor blades, position and velocity transformations are applied to the molecules as they move between stationary and rotating regions. The molecules acquire or lose the angular velocity of the rotating region as they enter or leave that region. The azimuth angle, relative to the fixed angle of the stationary region, of the rotating region at the time that the molecule crosses the boundary can be calculated and the position transformation adjusts for the change in azimuth angle. When in a rotating region, the molecule movement is subject to the appropriate centrifugal and Coriolis accelerations.

4.4 Sampling of flow properties

In order to calculate the macroscopic flow properties, the molecular properties are sampled at regular time intervals. The quantities that are sampled in each "sampling cell" are:

$\sum N_p$ the number of simulated molecules of species p.

$\sum N_p''$ the weighted number of simulated molecules of species p.

$\sum u_p'', \sum v_p'', \sum w_p''$ the weighted sums of the velocity components of the simulated molecules of species p.

$\sum \left(u_p^2\right)'', \sum \left(v_p^2\right)'', \sum \left(w_p^2\right)''$ the weighted sums of the squares of the velocity components of the molecules of species p.

$\sum \varepsilon_{rot,p}''$ the weighted sum of the rotational energy of the simulated molecules of species p.

$\sum i_{vib,p}''^{\,m}$ the weighted sum of the vibrational levels in the vibrational mode m of the simulated molecules of species p. (There are separate sums for each mode.)

$\sum \varepsilon_{el,p}''$ the weighted sum of the electronic energy of the simulated molecules of species p.

The summations are made over a total of N_{samp} samplings. Weighting factors are an undesirable, but often necessary, option for flows with cylindrical, spherical or axial symmetry.

The **sample size** provides an indication of the statistical scatter associated with the results and is equal to

$$\sum N_p \qquad (4.32)$$

for species p or

$$\sum_{p=1}^{q} \left(\sum N_p \right) \qquad (4.33)$$

for a gas mixture comprised of q molecular species.

The **number fraction** of species p is

$$n_p / n = \sum N_p / \sum_{p=1}^{q} \left(\sum N_p \right) = \sum N_p'' / \sum_{p=1}^{q} \left(\sum N_p'' \right). \qquad (4.34)$$

The **number density** of the overall gas is

$$n = F_N \sum_{p=1}^{q} \left(\sum N_p'' \right) / \left(N_{samp} V_{cell} \right), \qquad (4.35)$$

where V_{cell} is the volume of the sampling cell.

The overall gas **density** is, from Eqn. (1.5),

$$\rho = n \sum_{p=1}^{q} \left(m_p \sum N_p'' \right) / \sum_{p=1}^{q} \left(\sum N_p'' \right) \qquad (4.36)$$

and the **velocity component** in the x direction is, from Eqn. (1.6),

$$u_0 = \sum_{p=1}^{q} \left(m_p \sum u_p'' \right) / \sum_{p=1}^{q} \left(m_p \sum N_p'' \right), \qquad (4.37)$$

with similar expressions for the other components. The x component of the **diffusion velocity** of species p follows from Eqn. (1.8) as

$$U_p = \sum u_p'' - u_0 \qquad (4.38)$$

with similar equations for the other velocity components and species.

The **translational temperature** is, from Eqn. (1.14),

$$T_{tr} = \left[\sum_{p=1}^{q} \left[m_p \left\{ \sum \left(u_p^2 \right)'' + \sum \left(v_p^2 \right)'' + \sum \left(w_p^2 \right)'' \right\} \right] - \right.$$
$$\left. - \sum_{p=1}^{q} \left(m_p \sum N_p'' \right) \left\{ u_0^2 + v_0^2 + w_0^2 \right\} \right] / \left\{ 3k \sum_{p=1}^{q} \left(\sum N_p'' \right) \right\}. \qquad (4.39)$$

Separate temperatures may be defined for the molecular species and for the three velocity components. The differences between the various temperatures provide a measure of the translational and species related nonequilibrium in the gas. The temperature based on the x components of velocity is

$$T_{tr,x} = \left[\sum_{p=1}^{q} m_p \left\{\sum (u_p^2)''\right\} - \sum_{p=1}^{q}\left(m_p \sum N_p''\right) u_0^2\right] \bigg/ \left\{k \sum_{p=1}^{q}\left(\sum N_p''\right)\right\} \quad (4.40)$$

and there are similar expressions for $T_{tr,y}$ and $T_{tr,z}$. The translational temperature of species p is

$$T_{tr,p} = m_p \left\{\sum (u_p^2)'' + \sum (v_p^2)'' + \sum (w_p^2)'' - u_0^2 - v_0^2 - w_0^2\right\} \bigg/ (3k) \quad (4.41)$$

and that based on the x components of velocity of species p is

$$T_{tr,x,p} = m_p \left\{\sum (u_p^2)'' - u_0^2\right\} \bigg/ k \quad . \quad (4.42)$$

The overall **rotational temperature** is

$$T_{rot} = (2/k)\left\{\sum_{p=1}^{q}\left(\sum \varepsilon_{rot,p}''\right) \bigg/ \left[\sum_{p=1}^{q} \zeta_{rot,p}\left(\sum N_p''\right) \bigg/ \sum_{p=1}^{q}\left(\sum N_p''\right)\right]\right\} \quad (4.43)$$

Rotation is assumed to be fully excited and ζ_{rot} is either 2 or 3. The rotational temperature for species p is

$$T_{rot,p} = (2/k)\left(\sum \varepsilon_{rot,p}'' \big/ \zeta_p\right) . \quad (4.44)$$

The vibrational level rather than the vibrational energy is stored for each simulated molecule and the levels are summed in the sampling procedure. The levels are based on the harmonic oscillator model for vibration. This assumes that the levels $i = 0,1,2,3,\ldots$ have equal energy steps of $k\Theta_{vib}$, where Θ_{vib} is the characteristic vibrational temperature. The average level $\bar{i}$ of a molecule can be calculated and the energy associated with this is $\bar{i}k\Theta_{vib}$. This energy may be compared with the theoretical vibrational energy per molecule of $k\Theta_{vib}/\{\exp(\Theta_{vib}/T)-1\}$, to give

$$T_{vib} = \Theta_{vib}\big/\ln\left(1 + 1/\bar{i}\right) . \quad (4.45)$$

The corresponding number of degrees of freedom of vibration is therefore

$$\zeta_{vib} = 2\bar{i}\ln\left(1 + 1/\bar{i}\right). \tag{4.46}$$

These equations relate to a single mode of a single species. The vibrational temperature of mode l of species p is therefore

$$T_{vib,l,p} = \Theta_{vib,k,p}/\ln\left(1 + \sum N_p''/\sum i_{vib,p}''^l\right). \tag{4.47}$$

The number of effective vibrational degrees of freedom of the mode is

$$\zeta_{vib,l,p} = 2\left(\sum i_{vib,p}''^l/N''\right)\ln\left(1 + \sum N_p''/\sum i_{vib,p}''^l\right) \tag{4.48}$$

and the total number of vibrational degrees of freedom of species p is obtained by summing over the m modes. *i.e.*

$$\zeta_{vib,p} = \sum_{l=1}^{m}\zeta_{vib,l,p} \tag{4.49}$$

The vibrational temperature of species p is

$$T_{vib,p} = \sum_{l=1}^{m}\left(\zeta_{vib,l,p}\,T_{vib,l,p}\right)\Big/\zeta_{vib,p}\;. \tag{4.50}$$

The average number of vibrational degrees of freedom of vibration is

$$\zeta_{vib} = \sum_{p=1}^{q}\left(\zeta_{vib,p}\sum N_p''\right)\Big/\sum_{p=1}^{q}\left(\sum N_p''\right) \tag{4.51}$$

and the overall **vibrational temperature** is

$$T_{vib} = \sum_{p=1}^{q}\left(T_{vib,p}\sum N_p''\right)\Big/\sum_{p=1}^{q}\left(\sum N_p''\right) \tag{4.52}$$

The sum over the electronic levels, or electronic partition function, of species p at temperature $T_{tr,p}$ can be written

$$Q_{el} = \sum_{n=0}^{\infty}g_n\exp\left(\varepsilon_{el,n,p}/\left\{kT_{tr,p}\right\}\right) \tag{4.53}$$

and the equilibrium electronic energy is

$$\frac{\overline{\left(\varepsilon_{el,p}\right)_0}}{kT_{tr,p}} = \frac{\sum_{n=0}^{\infty}g_n\left(\varepsilon_{el,n,p}/\left\{kT_{tr,p}\right\}\right)\exp\left(-\varepsilon_{el,n,p}/\left\{kT_{tr,p}\right\}\right)}{\sum_{n=0}^{\infty}g_n\exp\left(-\varepsilon_{el,n,p}/\left\{kT_{tr,p}\right\}\right)}.$$

An electronic temperature can therefore be written

$$T_{el,p} = \frac{\sum \varepsilon''_{el,p}}{k \sum N''_p} \frac{\sum_{n=0}^{\infty} g_n \exp\left(-\varepsilon_{el,n,p}/\{kT_{tr,p}\}\right)}{\sum_{n=0}^{\infty} g_n \left(\varepsilon_{el,n,p}/\{kT_{tr,p}\}\right)\exp\left(-\varepsilon_{el,n,p}/\{kT_{tr,p}\}\right)} \qquad (4.54)$$

and the effective number of electronic degrees of freedom of species p is

$$\zeta_{el,p} = 2\left(\sum \varepsilon''_{el,p}/\sum N''_p\right)/\left(kT_{tr,p}\right). \qquad (4.55)$$

The overall **electronic temperature** is

$$T_{el} = \sum_{p=1}^{q}\left(T_{el,p}\sum N''_p\right)/\sum_{p=1}^{q}\left(\sum N''_p\right). \qquad (4.56)$$

The effective number of degrees of freedom of species p is

$$\zeta_p = 3 + \zeta_{rot,p} + \zeta_{vib,p} + \zeta_{el,p} \qquad (4.57)$$

and the temperature of species p is

$$T_p = \left(3T_{tr,p} + \zeta_{rot,p}T_{rot,p} + \zeta_{vib,p}T_{vib,p} + \zeta_{el,p}T_{el,p}\right)/\zeta_p. \qquad (4.58)$$

The overall gas temperature is then

$$T = \sum_{p=1}^{q}\left(\zeta_p T_p N''\right)/\sum_{p=1}^{q}\left(\zeta_p N''\right). \qquad (4.59)$$

4.5 Sampling of surface properties

The sampling for the surface properties is made during a molecule move whenever a molecule strikes a solid surface. Separate samples are made over specified surface elements. In addition, separate samples must be made for the molecular species and, in most cases, also for the incident and reflected molecular properties. The velocity components are transformed from the **flow coordinates** (x,y,z) to **surface coordinates** (x''',y''',z''') based on the orientation of the surface element. The x''' axis is normal to the surface element and is directed towards the element. The y''' and z''' axes lie along the surface element and the ambiguity in their orientation may be resolved by setting the y''' axis normal to the x axis. The computational time duration Δt_{samp} of the sampling process must be recorded. The quantities that are sampled for each surface element are:

$\sum N_p$ the number of incident simulated molecules of species p.

$\sum N_p''$ the weighted number of incident molecules of species p.

$\sum \left(m_p u_{i,p}''' \right)''$, $\sum \left(m_p v_{i,p}''' \right)''$, $\sum \left(m_p v_{i,p}''' \right)''$ the weighted momentum sums of the incident molecules of species p.

$\sum \left(-m_p u_{r,p}''' \right)''$, $\sum \left(-m_p v_{r,p}''' \right)''$, $\sum \left(-m_p v_{r,p}''' \right)''$ the weighted momentum sums of the reflected molecules of species p.

$\sum \left(m_p \left\{ u_{i,p}^2 + u_{i,p}^2 + u_{i,p}^2 \right\} / 2 \right)''$ the weighted translational energy sums of the incident simulated molecules of species p. Note that the translational temperature is not affected by the transformation.

$\sum \left(-m_p \left\{ u_{r,p}^2 + u_{r,p}^2 + u_{r,p}^2 \right\} \right)''$ the weighted translational energy sums of the reflected simulated molecules of species p.

$\sum \varepsilon_{rot,i,p}''$ the weighted rotational energy sums of the incident simulated molecules of species p.

$\sum -\varepsilon_{rot,r,p}''$ the weighted rotational energy sums of the reflected simulated molecules of species p.

$\sum \varepsilon_{vib,i,p}''$ the weighted vibrational energy sums of the incident simulated molecules of species p.

$\sum -\varepsilon_{vib,r,p}''$ the weighted vibrational energy sums of the reflected simulated molecules of species p.

$\sum \varepsilon_{el,i,p}''$ the weighted electronic energy sums of the incident simulated molecules of species p.

$\sum -\varepsilon_{el,r,p}''$ the weighted electronic energy sums of the reflected simulated molecules of species p.

The above sampled quantities are sufficient for determining the number flux, pressure, shear stresses and heat transfer to the surface, but it is desirable to also directly sample the velocity and temperature slips at the surface. This process employs the following sums that are made, for each surface element, over the incident and reflected quantities and over all molecular species:

$\sum(1/|u'''|)''$ the weighted sum of the reciprocal of the magnitude of the normal velocity component.

$\sum(m/|u'''|)''$ the weighted sum of the molecular mass divided by the magnitude of the normal velocity component.

$\sum(mv'''/|u'''|)''$ the weighted sum of the product of the molecular mass times a parallel velocity component divided by the magnitude of the normal velocity component.

$\sum(m\{u^2+v^2+w^2\}/|u'''|)''$ the weighted sum of the product of the molecular mass times the square of the speed divided by the magnitude of the normal velocity component.

$\sum(m/|u'''|)''$ the weighted sum of the molecular mass divided by the magnitude of the normal velocity component.

$\sum(\varepsilon_{rot}/|u'''|)''$ the weighted sum of the rotational energy divided by the magnitude of the normal velocity component.

$\sum(\zeta_{rot}/|u'''|)''$ the weighted sum of the number of rotational degrees of freedom divided by the magnitude of the normal velocity component.

As with the flowfield properties, the **sample size** provides an indication of the statistical scatter associated with the results and is equal to

$$\sum N_p \tag{4.60}$$

for species p or

$$\sum_{p=1}^{q}\left(\sum N_p\right) \tag{4.61}$$

for a gas mixture comprised of q molecular species. A similar summation over the species can be made for all the flow properties and, for brevity, only the overall values of the other surface properties are defined

The total **number flux** to the surface element is

$$\dot{N}_p = \sum_{p=1}^{q}\left\{\sum(N_p)\right\}F_N/\left(S_{el}\,\Delta t_{samp}\right). \tag{4.62}$$

The pressure on the element due to the incident molecules is

$$p_i = \sum_{p=1}^{q} \left\{ \sum \left(m_p u_{i,p}'''\right)'' \right\} F_N / \left(S_{el} \Delta t_{samp}\right) \qquad (4.63)$$

and that due to the reflected molecules is

$$p_r = \sum_{p=1}^{q} \left\{ \sum \left(-m_p u_{r,p}'''\right)'' \right\} F_N / \left(S_{el} \Delta t_{samp}\right).$$

and the net **pressure** is

$$p = p_i + p_r. \qquad (4.64)$$

The net **shear stress** in the y''' direction is

$$\tau = \left[\sum_{p=1}^{q} \left\{ \sum \left(m_p v'''\right)'' \right\} + \sum_{p=1}^{q} \left\{ \sum \left(m_p v'''\right)'' \right\} \right] F_N / \left(S_{el} \Delta t_{samp}\right) \qquad (4.65)$$

and there is a similar expression for the stress in the z''' direction. Both the incident and reflected pressures are necessarily positive and the reflected shear stress is zero for diffuse reflection. There is often a problem with heat transfer, especially at low Knudsen numbers, because the incident and reflected heat transfers are opposite in sign and the net heat transfer may be very small or even zero. In fact, there may be a requirement to set the surface temperature to the "adiabatic surface temperature" at which the net heat transfer is zero.

The incident translational heat transfer is

$$q_{i,tr} = \sum_{p=1}^{q} \left\{ \sum \left(m_p \left\{u_{i,p}^2 + u_{i,p}^2 + u_{i,p}^2\right\}/2\right)'' \right\} F_N / \left(S_{el} \Delta t_{samp}\right), \qquad (4.66)$$

the incident rotational energy is

$$q_{i,rot} = \sum_{p=1}^{q} \left\{ \sum \varepsilon_{rot,i,p}'' \right\} F_N / \left(S_{el} \Delta t_{samp}\right) \qquad (4.67)$$

and there are similar expressions for the incident vibrational and electronic heat transfer. In addition, there are similar expressions for all the reflected components of heat transfer. The net **heat transfer** is

$$q = q_{i,tr} + q_{i,rot} + q_{i,vib} + q_{i,el} - q_{r,tr} - q_{r,rot} - q_{r,vib} - q_{r,el} \qquad (4.68)$$

The continuum "zero slip" boundary conditions break down at finite Knudsen numbers and the velocity and temperature slips may be determined from the surface samples. The gas immediately adjacent to a surface is comprised of molecules with a positive value of

u''' that are about to collide with the surface and molecules with a negative value of u''' that have just collided with the surface. In both cases, the time that the molecules spend in an indefinitely thin layer adjacent to the surface is proportional to $|1/u'''|$ and their contribution to the spatial gas properties is proportional to this quantity alone.

The **velocity slip** in the y''' direction is therefore

$$v'''_{slip} = \sum \left(mv'''/|u'''| \right)'' / \sum \left(m/|u'''| \right)'' - V_{surf}, \tag{4.69}$$

where V_{surf} is the surface speed in the y''' direction.

The translational **temperature slip** is, similarly,

$$T_{tr,slip} = \frac{\sum \left(m\{u^2 + v^2 + w^2\}/|u'''| \right)'' - \sum \left(m/|u'''| \right)'' \left(v'''^2_{slip} + w'''^2_{slip} \right)}{3k \sum \left(1/|u'''| \right)''} - T_{surf} \tag{4.70}$$

and the rotational temperature slip is

$$T_{rot,slip} = (2/k) \sum \left(\varepsilon_{rot}/|u'''| \right)'' / \sum \left(\zeta_{rot}/|u'''| \right)'' - T_{surf}. \tag{4.71}$$

In addition to the sampling over the surface elements in order to determine the property distributions, it may be necessary to sample the overall forces and moments on a body that is immersed in the flow. With (x_c, y_c, z_c) as the location of the collision of the molecule with the body, the following sums are made:

$\sum \{ m(u_i - u_r) \}$, $\sum \{ m(v_i - v_r) \}$, $\sum \{ m(w_i - w_r) \}$ the momentum transfer in the x, y, z directions.

$\sum \left[m\{ (u_i y_c - v_i x_c) - (u_r y_c - v_r x_c) \} \right]$ the angular momentum transfer about the z axis.

$\sum \left[m\{ (w_i x_c - u_i z_c) - (u_r x_c - u_r z_c) \} \right]$, the angular momentum transfer about the y axis.

$\sum \left[m\{ (v_i z_c - w_i y_c) - (v_r z_c - w_r y_c) \} \right]$, the angular momentum transfer about the x axis.

The **force** on the body in the x direction is

$$F_x = \sum \{m(u_i - u_r)\}/\Delta t_{samp} ; \qquad (4.72)$$

with similar expressions for the components in the y and z directions.

The **moment** in the x-y plane about the x axis is

$$M_x = \sum \left[m\{(u_i y_c - v_i x_c) - (u_r y_c - v_r x_c)\} \right] \Big/ \Delta t_{samp} \qquad (4.73)$$

and is clockwise when looking in the negative x direction. Similar expressions can be written down for the moments about the other two axes. More complex sampling could be made to determine the relative contributions of pressure and shear stress to the forces and moments on the body.

4.6 Geometrical considerations

The space in which the molecules move is divided into a network of collision and sampling cells. These facilitate the selection of collision partners and the ordered output the gas properties over the flowfield. Early DSMC programs employed a single set of cells but, because typical applications now employ millions of simulated molecules, the desirable number of molecules per cell with regard to the sampling and output of the flow properties is several orders of magnitude greater than the desirable number for the selection of collision partners.

In most cases, the cells form a grid or mesh and may appear similar to, or may even be identical to, the grids and meshes that are employed in continuum CFD. However, continuum CFD grids are subject to mathematical constraints that do not apply to DSMC collision cells where the sole objective is to minimize the physical separation of the collision pairs. The purpose of DSMC sampling cells is to provide an acceptable representation of the flowfield properties and, although some procedures involve a sampled flowfield property, they have very little influence on the accuracy of the calculation.

In order to select potential collision partners from the same collision cell, it is necessary to index the simulated molecules to the cells. In the case of the regular rectangular cells in Fig. 4.1, given the coordinates of the location to which a molecule has moved, it is a simple matter to analytically determine the cell in which the molecule lies. The extension of a grid scheme from two dimensions, as shown in

Fig. 4.1, to the corresponding scheme in three dimensions should ideally increase the computational effort in the ratio 3:2. The regular rectangular cells satisfy this condition of linear scalability. Also, it is a simple matter to subdivide each cell into rectangular sub-cells such that the sub-cells are effectively collision cells and the larger cells are the sampling cells. This grid system can provide good results if the density variation over the flowfield is moderate and all surfaces lie along the cell boundaries. However, for the curved surface shown, the results will be unacceptable unless the Knudsen number is large.

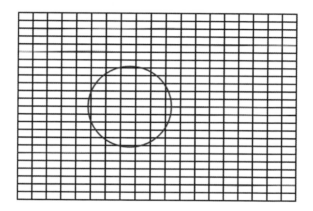

Fig. 4.1 A cylindrical surface embedded in a regular rectangular grid.

A good calculation for the flow past a cylinder can be made through the employment of a custom grid in cylindrical coordinates such as that in Fig. 4.2. This illustration is indicative only in that a real grid would employ a vastly larger number of cells. An analytical determination of the cell is still possible and it is possible to set cells in arithmetic progression and to have piecewise discontinuities in cell size. For a sphere, there could also be cells in the azimuthal direction and this has been implemented in a number of programs. However, it is far preferable to take advantage of the axial symmetry and, after the molecule move in three dimensions, the coordinates are rotated back to zero azimuth angle. The velocity components must also be rotated and the equations for the transformation have been presented as Eqns. (4.29-30).

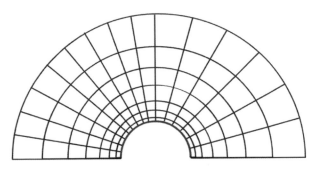

Fig. 4.2 A custom grid for the flow past a cylinder or sphere.

Custom grids were employed in the early days of DSMC but were limited to simple geometries and it became necessary to develop general purpose grids that could be applied to arbitrary geometries. One approach was to divide the flowfield into multiple zones with two straight and two curved sides. Judgment was required because the zones and cells had to match the flow that developed. This sometimes required a preliminary calculation and it was hardly possible to repeatedly adapt the cells in a continuing unsteady flow. As the geometries of interest became more complex, the effort became excessive. A further problem was that it was not possible to analytically determine the cell from the coordinates and it was necessary to calculate the intersections of the molecule trajectories with all cell boundaries. Sub-cells were introduced while these schemes were current and these exacerbated both problems.

A number of DSMC programs employ the "tree-structure" of cells that is illustrated in Fig. 4.3. The rectangular cells of Fig. 4.1 are regarded as zero level cells and are selectively divided through equal parts to some higher level (four in Fig. 4.3) of subdivision. The selection of cells to be subdivided to the next higher level is based on the expected number density and the proximity to a surface. In the case of cell adaption, it may be the sampled rather than the expected number density. The reduction in cell size at each level of subdivision is by a factor of two in a one-dimensional flow, four in a two-dimensional flow and eight in a three-dimensional flow. A factor of two reduction in cell volume is a reasonable value and a one-dimensional version of this scheme has been implemented for cell adaption of 1-D flows in the **DSMC.F90** program.

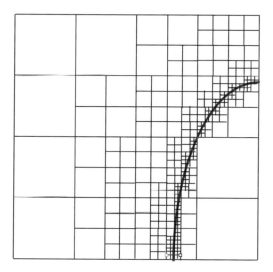

Fig. 4.3 A "tree-structured" grid in two dimensions.

A factor of four reduction in cell volume when moving to the next level of subdivision in a two-dimensional flow is acceptable as far as sampling cells are concerned. However, with the optimum number of molecules in a collision cell being about ten, factor of four steps in the sizes of the collision cells is too large. The factor of eight steps in three-dimensional versions of this scheme are unacceptable for either sampling or collision cells.

The finite volume grids that have been developed in the context of continuum CFD and solid mechanics have also been employed in DSMC programs. The simplest grids employ triangles in two dimensions and tetrahedra in three dimensions. Tetrahedral cells interface seamlessly with general three-dimensional surfaces that must be defined sets of triangular segments. A single set of triangular or tetrahedral cells for both collisions and sampling will inevitably lead to a poor DSMC simulation. However, if the molecules are indexed to an overlaid rectangular grid of collision cells, generally called transient sub-cells, that encompass the triangular or tetrahedral cell, a satisfactory result can be obtained. These are irregular grids or meshes and it is necessary to not only determine all collisions with cell boundaries, but also build extensive connectivity databases. There are many other classes of grid or mesh and some are regular in that connectivity is available through rows and columns.

The essential requirement of any DSMC cell scheme is that it should provide solutions that, for a given total number of simulated molecules, are no worse than those from competing cell schemes. Any scheme that cannot be readily adapted to the developing flow will be automatically ruled out. The cell schemes that satisfy this condition should then be ranked in terms of usability. The user must specify the geometry of the surfaces and the overall size of the flowfield, but the grid generation within these boundaries should ideally be transparent to the user. Moreover, it should not require recourse to external programs, especially if those programs require any significant degree of expertise and/or expense. This latter requirement disadvantages any program that depends on continuum meshes.

The cell scheme that has been developed for the **DS2V** and **DS3V** programs has arguably proved to be superior to any other scheme that has been implemented in DSMC. The flowfield, either two or three dimensional, is divided into regular **divisions** that are shown as black lines in Fig. 4.4 and **elements** that are shown as blue lines. The divisions approximate to sampling cells in the freestream or reference gas and the number of elements in a division is larger than the initial number of simulated molecules in the division. The collision and sampling cells structures are independent of one another. The location of a cell is defined by the coordinates of the **cell node** and the cell is comprised of the elements that are nearer to that node than to any other node. The number of cell nodes in a division is proportional to the number of molecules in the division and the nodes are placed at random, but with a near uniform spacing.

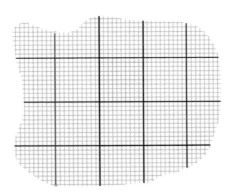

Fig. 4.4 Flowfield divisions and elements.

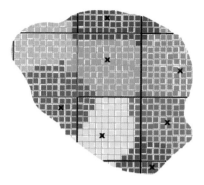

Fig. 4.5 Typical "nearest elements to the node" cells.

The cell formation process is shown in Fig. 4.5. Note that, when selecting the nearest elements, it is necessary to exclude any elements that have a surface intersection between the centre of the element and the cell node. The cells may be adapted to the flow such that there is very nearly the same number of simulated molecules in each cell. The specification of desired numbers of molecules in the separate collision and sampling cells is the only input that is required of the user. Moreover, the adaption of the cells to the flow is so rapid that it is feasible, in a flow with continuing unsteady sampling, to adapt the cells at every output interval. A further advantage of this scheme is that it scales linearly from two dimensions to three dimensions.

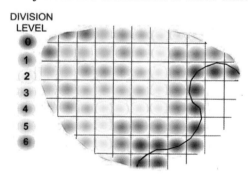

Fig. 4.6 Division levels relative to solid surfaces.

The surfaces are comprised of a large number of straight line or triangular segments and checks for surface interactions should only be made when necessary. The **DS2V/DS3V** programs define division **levels** relative to surfaces as shown in two dimensions in Fig. 4.6.

Divisions that contain surface segments are designated as level 0 divisions and those that are completely inside a surface are level -1. The divisions that have one or more corners in contact with a level 0 division are level 1, divisions that have one or more corners in contact with a level 1 division are level 2, and so on. The surface segments are indexed to the level 0 divisions and, during a molecule move step, it is only necessary to check for an impact on a surface segment when the molecule is near that segment.

The cell generation process may be illustrated by the cells that are generated for the second 2-D flow example in §8.3. This is for the Mach 10 flow of argon past a circular cylinder at a Knudsen number based on the diameter of 0.01. The cell nodes are initially placed at the centres of the divisions and, apart from those near the surface of the cylinder, the initial cells that are shown in Fig. 4.7 are near rectangular. The automatically set division array was 108×51 and the initial number of sampling cells was 4906. The initial number of molecules was 216,000 and there were 19,616 collision cells.

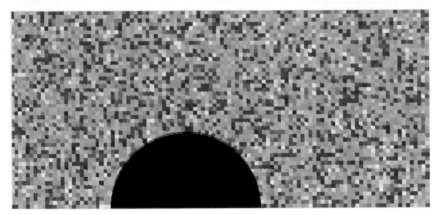

Fig. 4.7 A typical set of initial sampling cells.

The cells were adapted after steady flow had been attained and the resulting sampling and collision cells are shown in Figs. 4.8 and 4.9, respectively. The number of simulated molecules in the steady state averaged 272,000 and the adaption was to 8 molecules per collision cell and 27 molecules per sampling cell. The number of adapted sampling cells was 10,068 and the number of adapted collision cells was 33,964. Similar calculations were made with 8 million simulated molecules and there were then a million collision cells.

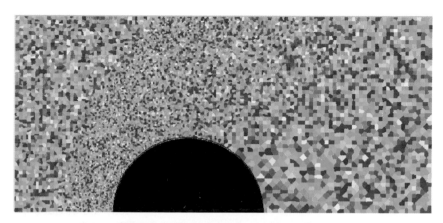

Fig. 4.8 The adapted sampling cells.

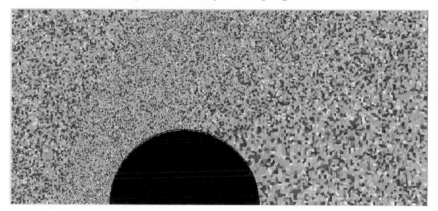

Fig. 4.9 The adapted collision cells.

References

Bird, G. A. (1994). *Molecular Gas Dynamics and the Direct Simulation of Gas Flows,* Clarendon Press, Oxford.

Chapman, S. and Cowling, T.G. (1970). *The Mathematical Theory of Non-uniform Gases* 3rd edn., Cambridge University Press.

Garcia, A. L. and Wagner, W. (2006). Generation of the Maxwellian Inflow Distribution, *J. Comput. Phys.* **217**, 693-708.

5

THE **DSMC.F90** PROGRAM

5.1 General description

The first version of the **DSMC.F90** program can be applied to **homogeneous gases** and **one-dimensional flows** with **plane, cylindrical** or **spherical** geometry. Future versions will extend it to cover **two-dimensional** flows with **plane** or **axially-symmetric** geometry and **three-dimensional** flows. All DSMC calculations have physical time as a variable and are **unsteady** flows. However, if the boundary conditions are such that a time-invariant state is attained at large times, this **steady** flow can be studied.

As indicated by the file subscript, it is a free form Fortran code and it does not contain any mixed-language statements. It has been developed under Intel Visual Fortran composer as a "console application". Upon execution, a Command Prompt window appears and the user is asked to specify whether it is to be a new run or the continuation of a previous run that has generated the necessary restart files. Should it be a new run, the user must specify whether it is for a homogeneous gas, a one-dimensional flow, a two-dimensional flow, an axially-symmetric flow, or a three-dimensional flow. The choice requires the presence of the appropriate data file that specifies the problem. There is a chapter devoted to each of these classes of flow and the first section of each chapter is concerned with the specification of the relevant data file.

The first items in the data files are the two integers that indicate the program version for which the data file was written. This allows backward compatibility such that a new version of the program does not force changes to data files that were written for earlier versions. The next item is the number of megabytes of storage that are required for the initial number of simulated molecules in the calculation. This is the only "computational variable" that is required in the data files and the number of simulated molecules must be sufficiently large to permit the criteria for a good DSMC simulation to be met.

The program can check whether the total number of simulated molecules is sufficiently large and warnings can be generated if the criteria for a good calculation have been met. However, it is not possible to make the program completely foolproof because the user is responsible for setting appropriate values for the overall size of the flowfield and for specifying the nature of the flowfield boundaries. It does not appear to be possible to enable the program to automatically detect whether or not the chosen boundaries are appropriate. The user, by setting the total number of sampling cells, controls the resolution of the output. A run-time option determines whether the output is for a continuing unsteady flow or an eventual steady flow.

The remaining data items specify the stream conditions, the shapes of the solid surfaces, and the nature of the boundary conditions. **All data items must be set in base SI units and the program employs base SI units throughout.** The data files do not fully specify the gaseous medium. Instead, the details of the gas species, or combinations of gas species and the potential chemical reactions between them, are set as subroutines within the source code. Section 5.2 provides a full specification of a gas subroutine and the user is expected to prepare a new subroutine whenever the existing subroutines are inadequate for the problem to be studied.

The program employs spatial cells that are generally irregular and are generated automatically by the program. Similarly, the time steps generally vary across the flowfield and the magnitudes of these steps are set within the program. However, the default choices for the grid size and time step may be changed within the source code which contains a designated section for the setting of **adjustable computational parameters**. The cell size is set through the number of simulated molecules within each computational cell. There are separate "collision" and "sampling" cells and the specified number of molecules in the sampling cells is generally at least one order of magnitude larger than the number in a collision cell. All calculations start at zero time and the initial cell structure reflects the initial state of the flow which is generally a uniform flow. The number of simulated molecules at any location generally changes with time and there is often a need to adapt the cells to reflect the changes in the distribution of number density over the flowfield. There is an opportunity for the user to initiate cell adaption whenever the flow is restarted.

5.2 Specification of a gas data subroutine

The data that describes the molecular species is set within the source code rather than as part of the data files. There are separate **gas data subroutines** and the data files contain a code number **IGAS** that specifies the subroutine that is to be used. Should a new gas model be required, a new subroutine must be added to the source code and a reference to this subroutine, through a new code number, must be added to the **READ_DATA** subroutine.

The first items in the subroutine are those that set the dimensions of the subscripted variables. Most of these deal with items that apply only to the more complex gas models that involve chemical reactions, but a value is required in every subroutine because all variables are dimensioned in the **ALLOCATE_GAS** subroutines. The default list from the hard sphere gas (**IGAS** = 1) can be copied to the new subroutine and changes made only where necessary. The complete list is:

MSP   the number of molecular species.
MMRM   0 if all molecular species are monatomic, or
             1 if one or more species have rotational degrees of freedom.
MMVM   the maximum number of vibrational modes of any species.
MELE   the maximum number of electronic levels of any species.
             The minimum is 1 because there is always the ground state.
MVIBL   the maximum number of vibrational levels to dissociation.
MEX   the number of exchange or chain reactions.
MMEX   the maximum number of chain or exchange reactions that involve the same pair of pre-collision molecular species.
MNSR   the maximum number of surface reactions.

There must be a call to the **ALLOCATE_GAS** subroutine after the specification of these variables and before the specification of the species and reaction variables. The subscript **L** in the following variables refers to the code number of a particular molecular species.

SP(1,L)   the diameter of the VHS model at the reference temperature.
SP(2,L)   the reference temperature.
SP(3,L)   the power law of the viscosity-temperature variation.
SP(4,L)   the reciprocal of the VSS scattering parameter (1 for VHS).
SP(5,L)   the molecular mass.
SP(6,L)   the heat of formation of a molecule at 273K.

This need be specified only if the species is involved in a chemical reaction.

ISPR(1,L) the number of rotational degrees of freedom.

The following items are required only if ISPR(1,L) > 0.

ISPR(2,L) 0 for a constant rotational relaxation collision number, or 1 for a polynomial temperature dependence of this number.

SPR(1,L)   the constant rotational relaxation collision number, or the constant in a second order polynomial in temperature.

The following two items are required only if ISPR(2,L) = 1.

SPR(2,L)   the coefficient of temperature in the polynomial

SPR(3,L)   the coefficient of temperature squared in the polynomial

ISPV(L)   the number of vibrational modes of molecular species L

The following items are repeated for each of the K *modes.*

SPVM(1,K,L) the characteristic temperature of the mode

SPVM(2,K,L) the reference value of the vibrational relaxation collision number.

SPVM(3,K,L) the temperature at which the reference value applies in the Millikan-White Eqn. (3.34), or -1 for SPVM(2,K,L) as a constant Z_v .

The program calculates the characteristic dissociation temperature from the heats of formation and sets it as SPVM(4,K,L).

SPVM(5,K,L) a constant reduction factor that is applied to the rate for the dissociation of mode K. This is set to unity unless there is compelling evidence for a reduction factor.

ISPVM(1,K,L) the species code of the first dissociation product.

ISPVM(2,K,L) the species code of the second dissociation product.

The items on the electronic levels are required only if MELE > 1 *and the electronic energy is to be taken into account.*

NELL(L)   the number of electronic levels that are included.

The following items are repeated for each of the J *levels.*

QELC(1,J,L) the degeneracy of the level

QELC(2,J,L) energy of the level divided by the Boltzmann constant.

QELC(3,J,L) the ratio of the excitation cross-section to the elastic cross-section.

Dissociation is regarded as an integral part of vibrational excitation and no additional data is required. The following data items relate to recombination reactions.

> When a data item relates to a pair L, M of molecular species, only the variable with L ≤ M need be entered. The corresponding data item with L > M is automatically set by the program.

ISPRC(L,M)   the species code of the recombined molecule that may result from a collision between species **L** & **M** molecules.

ISPRK(L,M)   the relevant vibrational mode of the recombined species.

SPRC(1,L,M,N)   the constant a in the ternary collision volume with species **N** as the third body molecule.

SPRC(2,L,M,N)   the temperature exponent b in the ternary collision volume with species **N** as the third body molecule.

SPRT(1,L,M)   the lower temperature value for the lists of cumulative dissociation probabilities.

SPRT(2,L,M)   the upper temperature value for the lists of cumulative dissociation probabilities.

(The program generates tables of vibrationally-resolved dissociation probabilities so that the vibrational level of the recombined molecule can be set for detailed balance. There is linear interpolation for gas temperatures between these limits that are set as data items.)

The final group of gas data items relates to exchange and chain reactions. **MEX** is the total number of distinct reactions and each pair **L,M** of molecular species may have up to **MMEX** possible reactions.

NSPEX(L,M)   the number of exchange or chain reactions associated with this pair of molecular species.

*In the following data items, **J** is 1 to **NSPEX(L,M)** and data is set for pairs of endothermic and exothermic reactions. (Here, the "forward reaction" is endothermic when writing the data for the endothermic reaction and exothermic when writing the data for the exothermic reaction.)*

NEX(J,L,M)   the code number(1 to **MEX**) of the reaction

ISPEX(J,1,L,M)   the code of the species that splits in the forward reaction.

ISPEX(J,2,L,M)   the code of the other species in the forward reaction.

ISPEX(J,3,L,M)   the code of the species that splits in the reverse reaction.

ISPEX(J,4,L,M)   the code of the other species in the reverse reaction.

ISPEX(J,5,L,M)   the vibrational mode of the species that splits in the forward reaction.

ISPEX(J,6,L,M)   the degeneracy of this reaction (the mode that splits).

ISPEX(J,7,L,M)   the vibrational mode of the species that splits in the reverse reaction. (Exothermic only)

SPEX(1,J,L,M)   the constant a in the reaction probability for the reaction. (Exothermic only)

SPEX(2,J,L,M)   the temperature exponent b in the reaction probability for the reaction. (Exothermic only)

SPEX(4,J,L,M)	the lower temperature value for the lists of cumulative dissociation probabilities to be applied in the reverse reaction. (Exothermic only)
SPEX(5,J,L,M)	the higher temperature value for the lists of cumulative dissociation probabilities to be applied in the reverse reaction. (Exothermic only)
SPEX(6,J,L,M)	the energy barrier to the reaction (separate values for the endothermic and exothermic reactions)

5.3 Adjustable computational parameters

The following parameters are located in the source code.

NMCC	(default 15) the initial number of molecules in a collision cell. The default number is 15, but a smaller number may be set in order to reduce the ratio of the mean separation of the molecules in a collision to the mean free path. It will be shown in §6.1 that the number can as low as 6. Larger values are automatically chosen in homogeneous gas calculations.
CPDTM	(default 0.2) The fraction of the local mean collision time that is the desired maximum time step.
TPDTM	(default 0.5) The fraction or multiple of a sampling cell transit time that is the desired maximum time step.
NMC	(default 0) Set to 0 if the collision partners are to be chosen at random from those in the same collision cell. Set to 1 if the collision partner is the nearest neighbour in the collision cell.
SAMPRAT	(default 5) The number of molecule move/collision steps between successive samples of the flow properties.
OUTRAT	(default 50) The number of samples between successive outputs of results.
FRACSAM	(default 0.5) In an unsteady flow, the output is sampled over a fraction the preceding output interval. This parameter specifies the fraction of this interval over which a sample is made. For example, if **OUTRAT** is 50 and **FRACSAM** is 0.2 the output will be based on an average over the final 10 samples.

ISAD (default 0) Set to 0 if cells are to be adapted only as a restart option. Set to 1 if the cells are to be adapted at every output interval in an unsteady flow (*not yet implemented*).

IMTS (default 0) Set to 0 if the move/collision time step is set to the overall time step that changes with time.
Set to 1 to use a cell dependent time step.
Set to 2 to keep the time step fixed at the initial value.

FNUMF (default 1) The setting of the number of real molecules that are represented by each simulated molecule assumes that the flowfield is a uniform stream or reference gas. This factor may be applied to this estimate when it is known that the flow density will depart markedly from the reference gas.

KREAC (default 1) Set to 1 to enable chemical reactions.
Set to 0 to disable chemical reactions.

TLIM (default 10^{20}) The default value effectively leads to an indefinite computation. A finite value causes the calculation to stop when the physical flow time reaches that value.

5.4 Auxiliary programs and post-processing

The input/output for the all-Fortran **DSMC** program is entirely through files and auxiliary programs are required for the generation of the more complex data and geometry files and for the generation of graphical output.

There are separate data files for zero (homogeneous gas), one two and three-dimensional flows. The number of items in a zero-dimensional case is so small that there is no need for a data generation program. For one-dimensional flows, the data file is larger and the **DS1D.DAT** data file can optionally be produced by the **DS1D.EXE** program. This executable program has been produced by the Xojo (formerly RealStudio and, previous to that, RealBasic) programming tool. This is a low cost program that can produce executables for Windows, Linux or OSX. Because Fortran compilers are available for all three operating systems, the programs are not restricted to Windows. The menu for the generation of a data file for one-dimensional flows is shown in Fig. 5.1.

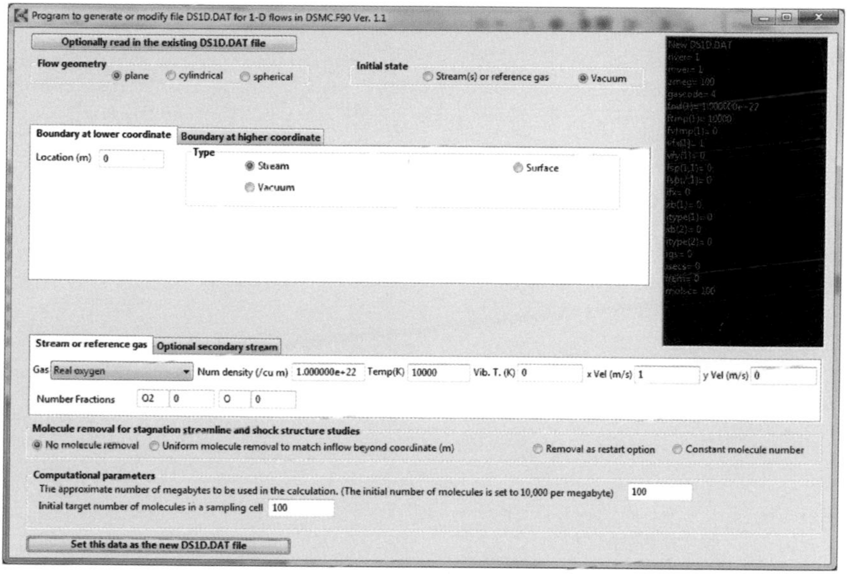

Fig. 5.1 The interactive menu screen for the generation of a **DS1D.DAT** file.

For two and three-dimensional flows, surface definition files as well as data files will be required. The surface definition in the two-dimensional case is in the form of a polyline with descriptive information on each segment. Interactive menu programs for both the general data and the surface data will be developed for two-dimensional flows. A similar data menu can be developed for three-dimensional flows, but the surface must triangulated and it will be necessary to have recourse to a commercial program such as Rhinoceros.

The output information from steady homogeneous cases is restricted to quantities such as collision rates or uniform surface properties. Unsteady flows require XY plots with flow or surface quantities plotted against time. The Tecplot (Version 9) program has been employed for all the XY plots in this book. This program does not require headers or special formatting for XY plots and the output routine in DSMC generates the ASCII files RELAX.DAT and/or COMPOSITION.DAT, with the variable list adjusted in the source code to suit the requirements of each application. These files are for post-processing and the comprehensive flow report files DS1OUT.DAT and, for continuing unsteady sampling, DS1n.DAT are always produced.

One-dimensional cases require XY plots for steady flows and two-dimensional contour plots in the x-t plane for unsteady flows. The data file for the XY plots is **PROFILE.DAT** and, again, the variable list may be modified in the source code. The data file for the contour plots is a formatted Tecplot file **DS1xt.DAT**. DSMC output is always for cells that apply to a finite volume of the flow. Tecplot expects data at points so that, for a plot to cover the entire flowfield, the data file must include extrapolated points on the sides and at the corners. Linear extrapolation has been employed and this can be problematic in some cases.

The 2-D and 3-D flow options have not yet been implemented in the **DSMC** code and the two-dimensional flows in this version of the book have been calculated with the **DS2** program. This is generally run inside the **DS2V** graphical shell program. **DS2V** is a Xojo program that contains the interactive data menu screens and also post-processing programs that report the surface and flowfield properties. Graphical representations of the surface and flowfield properties may be selected by the GUI program and data files may be output for XY plots by Tecplot and similar programs. Some versions of the program have also produced Tecplot data files. However, Tecplot can only deal with the irregular **DS2V** grid through triangulation and the results are often unusable. Boundary layer information is almost invariably distorted and triangles are often drawn outside the flowfield. More accurate information is provided by the post-processing that is built into **DS2V** and this has been used for most of the figures in Chapter 8.

Auxiliary programs will be written for the similar production of flowfield bitmaps of 2-D contours from **DSMC** output files but, instead of the fixed and relatively low resolution in **DS2V**, there will be higher resolution and more options. Tecplot is unable to deal with irregular three-dimensional data and the principal flowfield output will be 2-D x-y bitmaps for selected values of z. Xojo includes OpenGL and the contours of surface properties are readily produced from the sampled values on the triangular elements.

The data files are simplified by the requirement to set the gas models within the source code. The Q-K model is employed for chemical reactions and some of the reaction data requires pre-processing. This is through the auxiliary program **QKrates**. It would be extremely difficult to apply the Q-K model if this program was not available.

6

HOMOGENEOUS GAS APPLICATIONS

6.1 Description of the DS0D.DAT data file

> All dimensioned data must be in **base SI units**.

> Variables starting with I, J, K, L, M, and N should be entered as integers and all other variables should be entered as reals.

> In addition to the items in the .DAT files, the user is prompted to supply runtime input in the command window of the **DSMC** program. The start items that are common to all cases are:-
> ICLASS   sets the case as a homogeneous gas, one-dimensional, two-dimensional flow, etc. application.
> ISF   sets whether it is an eventually steady flow or a continuing unsteady flow.

NVER   the n in the **DSMC.F90** version $n.m$ for which the file was generated.

MVER   the m in the **DSMC.F90** version $n.m$ for which the file was generated.

IMEG   approximate number of megabytes to be used by the program. (The initial number of simulated molecules will be 10,000 times IMEG)

IGAS   the code number of the gas data subroutine that is to be employed in the calculation. The current codes are:

1   $d = 4e\text{-}10$ m hard sphere [MSP=1, MMVM=0]
2   argon [MSP=1, MMVM=0]
3   ideal nitrogen [MSP=1, MMVM=0]
4   real oxygen [MSP=2, MMVM=1]
5   for ideal air [MSP=2, MMVM=0]
6   real air appropriate to re-entry speeds [MSP=5, MMVM=1]
7   helium-argon-xenon mixture [MSP=3, MMVM=0]
8   combustible hydrogen-oxygen mixture [MSP=8, MMVM=3]

> The gas specification subroutines set a larger number of descriptive variables. The variables **MSP** and **MMVM** are listed here only because data items are affected by their values.

FND   the initial number density of the gas.
FTMP   the initial gas temperature.
If **MMVM>0**,   **FVTMP**   initial vibrational and electronic temperature.
If **MSP>1**, *in a loop n over* **MSP**
   FSP*(n)*   the fraction of this species in the initial gas.
end of loop over molecular species.

6.2 Collision rates

Homogeneous gas calculations with **DSMC.F90** are actually made as one-dimensional calculations in a stationary gas between planes of symmetry separated by twenty mean free paths. There is only one sampling cell and the number of collision cells is a computational parameter that is set within the source code through the specification of the desired number of simulated molecules in a collision cell.

The first set of collision test calculations with **DSMC.F90** employed a data file that is reported in the data description file **DS0D.TXT** as:

```
Data summary for program DSMC
 The n in version number n.m is            1
 The m in version number n.m is            1
 The approximate number of megabytes for the calculation is
10
           1
Hard sphere gas
 The gas properties are:-
    The stream number density is  1.000000000000000E+022
    The stream temperature is   300.000000000000
```

The calculations employ a hard sphere gas and all calculations were made with 100,000 simulated molecules. The ratio of the collision rate to the equilibrium collision rate of Eqn. (2.33) was determined as the average number of molecules per collision cell rate was varied. Calculations were made with the default procedure which selects the molecules from anywhere within the collision cell and also with the option that selects nearest-neighbour collision pairs. The quality of the results improves as the ratio of the mean separation of the colliding molecules to the mean free path decreases. With the default procedure, this ratio becomes smaller as the average number of molecules per collision cell decreases. With nearest-neighbour collisions, it has its minimum value independent of the collision cell size but, because this option has on one occasion led to undesirable side effects, the minimum acceptable number of molecules per collision cell is a critical factor.

Table 6.1 The effect of collision cell size within the default collision procedure.

Mean molecules per collision cell	Collision rate / equilibrium rate	Mean coll. sep. / mean free path	Mean cell time / overall flow time
20	0.9994	0.0474	1.0000
15	0.9994	0.0355	1.0000
10	1.0003	0.0237	1.0000
8	1.0028	0.0190	1.0000
6	1.0170	0.0143	1.0000
5	1.0418	0.0118	0.9998
4	1.0525	0.0095	0.9562
2	1.0490	0.0048	0.6247

There is very good agreement between the computational and theoretical collision rates when the average number of simulated molecules per collision cell exceeds ten. Below this number, there is an increase in the rate ratio that has a maximum of just over five percent at an average molecule number of four. There is then a slight decrease in the rate, but the procedures have then broken down to such an extent that the collision cell times are unable to keep up with the overall flow time. The calculation employed the procedure of Eqn. (4.18) that involves only the instantaneous molecule number rather than the traditional procedure that employed the average number. The program was temporarily modified to use the traditional procedure, but there was no significant change in the collision rates. The calculation with an average of four simulated molecules was also repeated for the VHS model for argon and the excess rate of just over 5% was unchanged.

The ratio of the mean collisional separation to the mean free path, or mcs/mfp ratio is directly proportional to the average number of simulated molecules. It is therefore desirable to make the collision cells as small as possible, but it appears to be undesirable to go below an average of eight simulated molecules per collision cell.

Similar calculations were made with the nearest-neighbour selection procedure in place of the default procedure. The results for the collision rate were similar to those in Table 6.1, but there was a uniform reduction in the rate by about 0.08%. The mcs/mfp ratio was 0.00395 in all cases. This is almost one fifth the ratio with the default procedure and a mean number of eight simulated molecules per collision cell. There is a great deal to gain from nearest-neighbour collisions in low Knudsen number flows where the mcs./mfp ratio is the critical factor that determines the degree of convergence of the results.

All the calculations suppress successive collisions of a molecule with the same collision partner. This is achieved by storing the code number of the previous collision partner of each molecule. This memory of the previous collision partner is carried across the time steps. Most implementations of the conventional NTC method have made no attempt to eliminate duplicate collisions. To determine the magnitude of the possible errors, the calculations in Table 6.1 were repeated without suppression of duplicate collisions. The duplicate collisions were about one percent of the total collisions when there were about ten molecules per collision cell. This increased to two percent at four molecules per cell. On the other hand, without the suppression of duplicate collisions, the collision rate increased by about 0.05% and the agreement with the theoretical equilibrium rate is almost exact when there are ten or more simulated molecules per collision cell. Duplicate collisions typically reduce the effective collision rate by about 1% and, on balance, it appears desirable to eliminate the duplicate collisions and tolerate the much smaller real reduction in the collision rate. Note that the duplicate collisions in most existing calculations would have partly canceled the effect of the excessive high collision rate for small numbers of molecule per cell. Given the degree of breakdown in the procedures that is indicated by the significant lag in the collision cell times when there is an average of two molecules per collision cell, the breakdown has a remarkably small effect on the collision rate. The benign nature of the breakdown explains the good results that have generally been reported from DSMC calculations with unreasonably small numbers of simulated molecules in a collision cell.

The measurement of the collision rate ratios to four significant figures has been facilitated by the measures to minimize statistical scatter in the setting of the initial gas. These are described in §4.1. For example, with 100,000 simulated molecules, a typical temperature was 300.01 K and the nominally zero stream speeds were of the order of 0.001 m/s.

The next check was on the distribution of the collisions between the various molecular species in a gas mixture. The calculation again employed a total of 100,000 simulated molecules with an average of 15 per collision cell. The collision pairs were chosen from anywhere in the collision cell. The new data file was reported in **DS0D.TXT** as:

```
Data summary for program DSMC
  The n in version number n.m is          1
  The m in version number n.m is          1
  The approximate number of megabytes for the calculation is 10
  7    Helium-argon-xenon mixture
  The gas properties are:-
    The stream number density is  1.000000000000000E+022
    The stream temperature is    300.000000000000
    The fraction of species          1  is  0.100000000000000
    The fraction of species          2  is  0.700000000000000
    The fraction of species          3  is  0.200000000000000
```

The sampled number of species dependent collisions and the ratio of this number to the theoretical number that is predicted by Eqn. 2.36 are shown in Table 6.2.

Table 6.2 The sampled collision numbers and rates in a ternary gas mixture.

	Helium	Argon	Xenon
Helium number	154,333,660	1,543,644,182	654,312,329
ratio to equilib.	1.0032	1.0002	0.9967
Argon number	1,543,644,182	7,546,285,862	2,450,141,552
ratio to equilib.	1.0002	1.0007	0.9994
Xenon number	654,312,329	2,450,141,552	640,190,249
ratio to equilib.	0.9967	0.9994	0.9992

There has been concern (Bird, 1994) about the use of a single value of the product of cross-section and relative speed when very light molecules are present. However, the computation time for this calculation was only about 15% higher than that for the simple gas.

6.3 Relaxation processes

The procedures for molecules with internal degrees of freedom must lead to equipartition and must reproduce the nominal relaxation rate. The first test case is for ideal nitrogen which has two degrees of freedom of rotation, but is assumed to not have a vibrational mode. The calculation is for a case in which there is initially no energy in the rotational mode. The data file was reported in **DS0D.TXT** as:

```
Data summary for program DSMC
The n in version number n.m is          1
The m in version number n.m is          1
The approximate number of megabytes for the calculation is    1000
3    Ideal nitrogen
The gas properties are:-
    The stream number density is  1.000000000000000E+022
    The stream temperature is   500.000000000000
```

There is no data item that allows the initial rotational temperature to be set to zero. A temporary statement was therefore added to re-set the initial rotational energy of each molecule to zero. Ten million simulated molecules were employed and the "continuing unsteady flow" was chosen when the run was started. In addition, the default time step was to a small fraction of the default value so that the temperatures were reported in the **RELAX.DAT** file several times during each mean collision time.

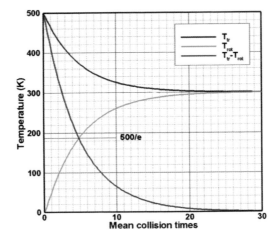

Fig. 6.1 Rotational relaxation in nitrogen with $Z_{rot} = 5$.

The rotational relaxation collision number Z_{rot} for nitrogen, as set in the gas definition subroutine in the source code, is a temperature independent value of 5. The Larsen-Borgnakke rotational redistribution was therefore applied to one in five collisions and, as shown in Fig. 6.1, the temperature difference fell to $1/e$ of its initial value at 4.95 mean collision times. The adequacy of this simple procedure has been questioned and (Haas et al, 1994) presented a far more complex relationship between Z_{rot} and the fraction of collisions that should be subject to rotational distribution. The case with $Z_{rot} = 5$ does not provide a good test because the two procedures are in close agreement for that case. A difference of about 10% is predicted for $Z_{rot} = 2$ and the results for this case are shown in Fig. 6.2. The sampled relaxation collision number was 2.06 and, given the uncertainties associated with relaxation time data, the simpler procedure is employed in the program

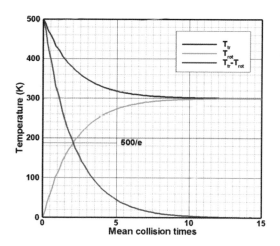

Fig. 6.2 Rotational relaxation in nitrogen with $Z_{rot} = 2$.

With regard to equipartition, there are three translational degrees of freedom and two of rotation, so that the initial nominal temperature of 500 K should fall to 300 K at equilibrium. The translational and rotational temperatures after 5,000 mean collision times in a similar calculation with one million simulated molecules were 300.14 K and 300.15 K respectively.

A similar test calculation was made for vibrational relaxation in oxygen. The data file was reported in **DS0D.TXT** as:

```
Data summary for program DSMC
  The n in version number n.m is          1
  The m in version number n.m is          1
  The approximate number of megabytes for the calculation is   100
          4
Real oxygen
The gas properties are:-
      The stream number density is  1.000000000000000E+022
      The stream temperature is   10000.0000000000
      The stream vibrational and electronic temperature is 0.000000
      The fraction of species         1  is  1.00000000000000
      The fraction of species         2  is  0.000000000000000E+000
```

Dissociation and electronic excitation were suppressed by the activation of internal switches within the source code. The initial temperature of 10,000 K was chosen because vibrational relaxation is so slow at lower temperatures that an excessively long computer run would have been required. Even so, the relaxation process in one million simulated molecules required about 3,000 mean collision times.

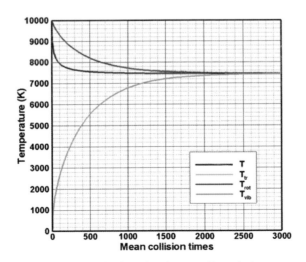

Fig. 6.3 Vibrational relaxation in non-dissociating oxygen without electronic modes.

Both the translational and rotational temperatures were initially set to 10,000 K and, because of the fast rotational relaxation they remain in equilibrium. The gas came to equilibrium at a temperature of 7,450 K. The number of vibrational degrees of freedom associated with the simple harmonic model of vibration in oxygen at 7,450 K is given by Eqn. (4.39) as $\zeta_{vib} = 1.712$. The number of degrees of freedom increases from 5 to 6.712 so that the expected value of the equilibrium temperature is 7,450 K and the results are consistent with equipartition.

The vibrational relaxation collision number changes rapidly with temperature and it is not clear from Fig. 6.3 whether the nominal rate has been achieved by the simple procedure of setting the fraction of collisions with vibrational adjustment to the inverse of the collision number. There is an option in the gas database for setting a fixed vibrational relaxation collision number and the calculation was repeated for a fixed $Z_{vib} = 200$ and the results are shown in Fig. 6.4.

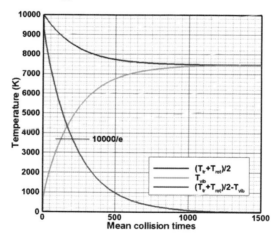

Fig. 6.4 Vibrational relaxation in oxygen with $Z_{vib} = 200$.

The difference between the mean of the translational and rotational temperatures and the vibrational temperature falls to $1/e$ of its original value at 206 mean collision times and the calculation again supports the use of the simpler procedure. The overall temperature at equilibrium was 7449 K, the translational 7449 K, the rotational 7460 K and the vibrational temperature was 7437 K. There are therefore discrepancies of the order of 0.2% that do not show in the plots.

The calculation that led to Fig. 6.3 was repeated with the inclusion of electronic excitation. There is no data for an "electronic relaxation collision number", but the number of electronic levels is small and data may be available on the excitation cross-sections of particular levels. The ratio of the elastic cross-section to the electronic excitation cross-section is equivalent to the relaxation collision number. The specification of the gas data subroutine requires data for each level, but equipartition appears to not be achieved unless they are equal. With this in mind and in the absence of data, the cross-section ratio for all the electronic levels in oxygen has been set to 500.

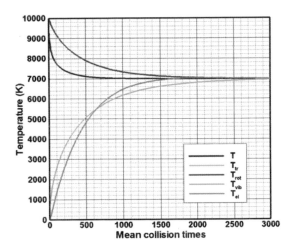

Fig. 6.5 Vibrational and electronic relaxation in non-dissociating oxygen.

The results in Fig. 6.5 provide verification of the procedures that have been developed for the treatment of the electronic modes in DSMC. The most striking result is that the inclusion of electronic excitation causes the equilibrium temperature to drop from 7450 K to 6980 K. This means that the equilibrium electronic energy in oxygen at about 7,000 K is equivalent to almost half a degree of freedom. There are no electronic levels in molecular nitrogen in the energy range of interest, but there is electronic energy in atomic nitrogen and atomic oxygen. The electronic modes have been almost universally neglected in both DSMC and Navier-Stokes calculations of very high temperature gas flows. The equilibrium electronic energy of a molecule at temperature T is

$$\frac{\varepsilon_{el}}{kT} = \frac{\sum_{n=0}^{\infty} g_n \left(\varepsilon_{el,n}/\{kT\}\right)\exp\left(-\varepsilon_{el,n}/\{kT\}\right)}{\sum_{n=0}^{\infty} g_n \exp\left(-\varepsilon_{el,n}/\{kT\}\right)}. \qquad (6.1)$$

This has been evaluated for the components of high temperature air and the results are shown in Fig 6.6.

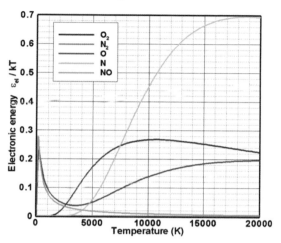

Fig. 6.6 Equilibrium electronic energy in high temperature air.

The plot confirms that the equilibrium electronic energy in molecular oxygen at 7,000 K is 0.48 degrees of freedom. The average energy over the components of high temperature air is less than this, but could well amount to half a degree of freedom at higher temperatures. While the relaxation rates for electronic energy are uncertain, radiation is observed immediately behind strong shock waves. It appears that a typical error in all the DSMC and continuum CFD calculations for re-entry flows that have ignored electronic energy would have been of the order of several percent.

6.4 Dissociation and recombination

A typical dissociation-recombination reaction can be represented as **AB+T↔A+B+T**, but the demonstration cases will be concerned with the dissociation **A₂+T↔2A+T** of symmetrical diatomic molecules. The degree of dissociation α is defined as the mass fraction of dissociated gas. Therefore, for the dissociation of a symmetrical diatomic molecule,

$$\alpha = n_A m_A / \rho = n_A / (n_A + 2n_{AA}).$$ (6.2)

In the case of the DSMC simulation of dissociation-recombination in a homogeneous gas, it is readily shown that this degree of dissociation is the ratio of the final to the initial number of simulated molecules minus one. The statistical mechanics result for the equilibrium degree of dissociation α_0 is

$$\frac{\alpha_0^2}{\alpha_0 - 1} = \frac{m_A \left(\pi m_A kT\right)^{3/2}}{2\rho h^3} \frac{\left(z_{rot}^A z_{vib}^A z_{el}^A\right)^2}{z_{rot}^{AA} z_{vib}^{AA} z_{el}^{AA}} \exp\left(-\frac{\Theta_{diss}}{T}\right).$$ (6.3)

This is plotted for oxygen as a function of temperature in Fig. (6.7) at standard density, 0.1 standard density and 0.01 standard density. The background material that is relevant to Eqns. (6.1) to (6.3) can be found in Vincenti and Kruger (1965).

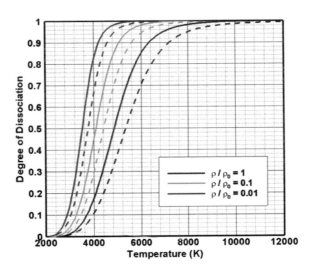

Fig. 6.7 The equilibrium degree of dissociation of oxygen. The full lines include electronic energy and the dotted lines are the corresponding results when electronic energy is neglected.

Note that, even when the electronic energy is excluded, it is important to include the degeneracies of the electronic ground states and to not just ignore the electronic partition functions.

The first test case is to follow the dissociation-process when molecular oxygen at standard density is initially at 15,000 K. The data is reported in **DS0D.TXT** as:

```
Data summary for program DSMC
  The n in version number n.m is          1
  The m in version number n.m is          1
  The approximate number of megabytes for the calculation is          100
  4
Real oxygen
  The gas properties are:-
    The stream number density is   2.686780000000000E+025
    The stream temperature is    15000.0000000000
    The stream vibrational and electronic temperature is 15000.0000
    The fraction of species          1  is     1.000000000000000
    The fraction of species          2  is    0.000000000000000E+000
```

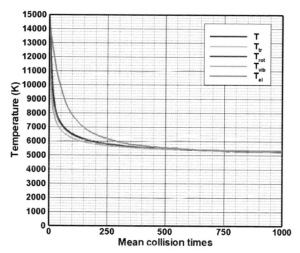

Fig. 6.8 Dissociation-recombination of oxygen initially at standard density and a temperature of 15,000 K.

The overall temperature remains in equilibrium with the translational and rotational temperatures throughout the relaxation process. This is because the slight lead in the vibrational temperature is matched almost exactly by a larger lag in the electronic temperature. In order to have comparable relaxation times for dissociation and electronic excitation, the ratio of the elastic cross-section to the electronic excitation cross-section has been set to 50. The equilibrium degree of dissociation is 0.645 at a temperature of 5,270 K. This is in agreement with the theoretical results from Eqn.

(6.3) that are plotted in Fig. (6.7). However, the agreement serves only to verify the coding because the coefficients a and b in Eqn. (3.34) for the ternary collision volume are also based on Eqn. (6.3). These coefficients form part of the gas specification subroutines and the auxiliary program **QKrates.exe** permits the interactive determination of the appropriate values.

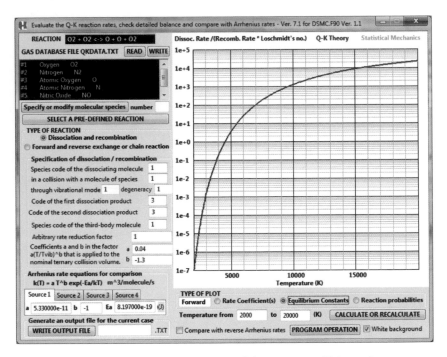

Fig. 6.9 Interactive determination of the ternary collision volume.

An image of the relevant screen that is generated by **QKrates.exe** for the dissociation-recombination reaction in oxygen is shown in Fig. 6.9. The interactive setting of the coefficients a and b to 0.04 and -1.3, respectively, leads to almost exact agreement between the Q-K and statistical mechanics equilibrium constants over the whole temperature range. This is for molecular oxygen as the collision partner or third body and the corresponding coefficients for reactions that involve atomic oxygen are 0.05 and -1.1.

It has been possible to determine the values of the ternary collision volume coefficients only because the Q-K criteria for dissociation and recombination within the DSMC collision procedures are so simple that analytical expressions can be written for the rate coefficients. Moreover, the dissociation rate coefficient is vibrationally resolved and this makes it possible to satisfy detailed balance by setting the vibrational distribution of the recombined molecules to be the same as the vibrational distribution of the dissociating molecules.

The procedures for the establishment of detailed balance are transparent to the user because the Q-K theoretical cumulative distribution function for the vibrational levels of the dissociating molecules is calculated within the **DERIVED_GAS_DATA** subroutine. Then, in the recombination procedures, a random fraction is generated and the vibrational level of the recombined molecule is set to the level that has this random fraction within its cumulative range. Figure 6.10 shows the result that corresponds to that in Fig. 6.7 when the detailed balance procedure is by-passed.

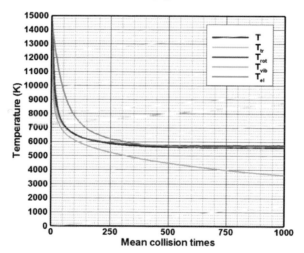

Fig. 6.10 Dissociation-recombination of oxygen when the detailed balance procedure is by-passed.

The vibrational temperature does not come to equilibrium with the other temperatures and the overall temperature tends to 5,600 K with the degree of dissociation equal to 0.627. Unlike the values when detailed balance is enforced, these values lie well off the theoretical curve in Fig. 6.7.

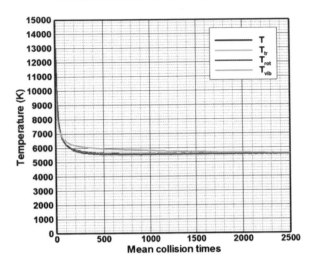

Fig. 6.11 Dissociation-recombination of oxygen when the electronic energy is neglected.

The test case was repeated with the enforcement of detailed balance, but with **MELE** set to unity in the gas data so that the electronic energy is not taken into account. A similar change must be made in the **QKrates** program in order to obtain new values of the a and b coefficients. These are 0.08 and -1.2 for dissociation in a collision with molecular oxygen and 0.16 and -1.05 for a collision with atomic oxygen. Figure 6.11 shows that the vibrational temperature converges to the overall temperature from above rather than below and equilibrium is attained far more slowly. The physical cause of these surprisingly large qualitative differences is unclear. However, the final degree of dissociation of 0.572 at 5,580 K is again in good agreement with the predicted result that is shown as the dashed curve for standard density in Fig. 6.7.

The QKrates.exe program also allows the reaction rates that are predicted by the Q-K theory to be compared with up to four rates in the form of the Arrhenius equation. Figure 6.12 shows good agreement with four of the commonly accepted dissociation rate coefficients. The "Source 1" Arrhenius rate is the rate that has been used for oxygen dissociation in the author's earlier programs that implemented the TCE reaction model. The agreement for this case is almost exact and the Q-K result generally lies within the uncertainty limits of the measured rates. In the cases where there does appear to be a significant disagreement, it is the Q-K rate that is the higher. There is, therefore, an option to apply a constant reduction factor to the Q-K rate. This factor will generally remain at its default value of unity and is unity for all the cases that are discussed in this section.

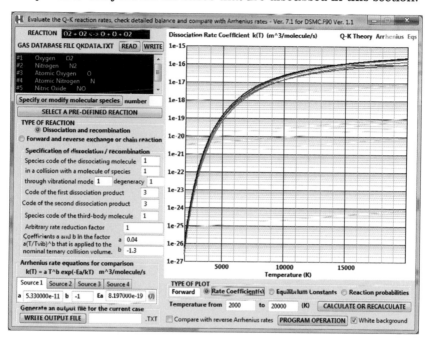

Fig. 6.12 A comparison of the Q-K rate coefficient for the dissociation of oxygen with four rates based on commonly used Arrhenius equations.

The Q-K condition for dissociation applies irrespective of whether the gas is in equilibrium. On the other hand, the derivation of the analytical result for the dissociation rate coefficient assumes equilibrium distributions of both translational velocities and vibrational levels. The derivation of the ternary collision volume that leads to an equilibrium degree of dissociation also assumes equilibrium translation and vibration. There will not be equilibrium if the collision number for the physical process, in this case dissociation, in much smaller than the vibrational relaxation collision number. The average number of collisions between dissociations is given by the Q-K theory as a function of temperature. For the dissociation of oxygen, this is compared with the vibrational collision number in Fig. 6.13.

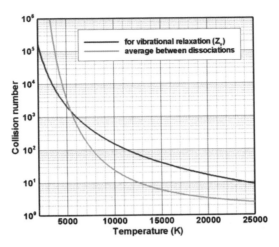

Fig. 6.13 A comparison of the dissociation and vibrational relaxation collision numbers in oxygen.

The two are comparable at temperatures around 5,000 to 6,000 K and this explains why equilibrium has been attained in the main test case shown in Fig. 6.7. On the other hand, the dissociation rate is about seven times faster than the vibrational relaxation rate at the initial temperature of 15,000 K and, as shown in Fig. 6.8, the temperatures are not in equilibrium in the first few nanoseconds of the dissociation process.

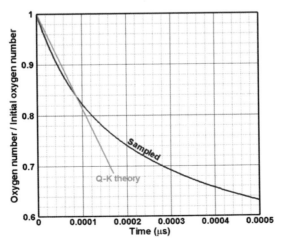

Fig. 6.14 The initial dissociation rate in standard density oxygen at 15,000 K

Because of the initial nonequilibrium, it is to be expected that the initial dissociation rate will not quite match the theoretical rate of Eqn. (3.38). This is confirmed in Fig. 6.14 where the sampled drop in the molecular oxygen number density is compared with the Q-K prediction.

6.5 Chain and exchange reactions

The Q-K chemistry model that is presented in this book has evolved through a number of versions that have involved different treatments of the reverse reactions. The recombination reaction in the initial presentation of the method (Bird, 2009) was set by a reaction probability that essentially amounted to a specified Arrhenius equation. This was later changed (Bird, 2011a) to a reaction probability based on the probability of a third molecule being within a "ternary collision volume" that was based on physical arguments. The current procedure sets this collision volume to a constant multiplied by a power of the temperature. The two constants can be set to satisfy the law of mass action through the equilibrium constant. The initial procedure for reverse or exothermic exchange and chain reactions employed an unsatisfactory procedure and (Bird, 2011a) evaluated the equilibrium constant for every reaction. The current procedure is similar to that for recombination in that the reaction probability that satisfies the law of mass action is defined as an expression that includes a constant multiplied by a power of the temperature. The current procedures are described in §3.6.

An alternative procedure for exchange and chain reaction was presented by Bird (2011b). This employed the forward or endothermic condition for both forward and reverse reactions and mass action was satisfied by an upward adjustment of either the forward or reverse activation energy from their default values. These are the negative of the heat of reaction for endothermic reactions and zero for exothermic reactions. Many of these adjustments turned out to be unphysical and the alternative procedure is no longer recommended. However, it was employed in a study (Bird, 2012b) that demonstrated the need to enforce detailed balance with regard to the pre-reaction and post-reaction vibrational levels. The test case in this section repeats these calculations with the procedures that are now recommended.

The data file for the test case is reported in **DS0D.TXT** as:

```
Data summary for program DSMC
  The n in version number n.m is          1
  The m in version number n.m is          1
  The approximate number of megabytes for the calculation is 1000
          8        Oxygen-hydrogen
  The gas properties are:-
      The stream number density is   1.000000000000000E+020
      The stream temperature is    3000.00000000000
      The stream vibrational and electronic temperature is
  3000.00000000000
          The fraction of species      2  is  0.500000000000000
          The fraction of species      3  is  0.500000000000000
```

The fractions of the other six species in the oxygen-hydrogen database were set to zero.

This is for the simulation of a homogeneous gas initially comprised of equal parts by number of oxygen and atomic hydrogen at 3,000 K and a number density of 10^{20} m^{-3}. All reactions other than $O_2+H \leftrightarrow OH+O$ are suppressed by special coding in the gas definition file. Dissociation and recombination are temporarily by-passed in **DSMC.F90**. The energy barriers have their default values of zero.

The cumulative distribution function for the vibrational levels of the pre-reaction oxygen molecules is calculated within the program from Eqn. (3.37) that is based on the Q-K theory. This distribution is applied to the oxygen molecules that are produced by the reverse reaction in order to satisfy the detailed balance principle. If detailed balance is not enforced, the post reaction distribution of oxygen vibrational states is set by the Larsen-Borgnakke procedures to an equilibrium distribution. The detailed balance distribution from the Q-K theory at 2,000 K and 3,000 K is compared with the corresponding cumulative equilibrium distribution in Table 6.3

The activation energy for the reverse reaction is zero and the pre-reaction vibrational distribution of OH is the same as that in non-reacting collisions. However, equilibrium is not achieved if the non-reacting Larsen-Borgnakke procedures are employed to set the post-reaction vibrational level of OH. This is because the available energy for distribution to vibration is reduced by heat of reaction. The post-reaction vibrational levels are, instead, selected from the equilibrium distribution at the overall temperature. This is complicated by the need to avoid negative collision energies. Similar considerations apply to the selection of post-reaction rotational energies.

Table 6.3 Comparison of the Q-K and Boltzmann cumulative distributions.

Vibrational level	2,000 K		3,000 K	
	Q-K	Boltzmann	Q-K	Boltzmann
0	0.3004	0.6763	0.2724	0.5286
1	0.5526	0.8952	0.5049	0.7778
2	0.7525	0.9661	0.6946	0.8952
3	0.8939	0.9890	0.8368	0.9506
4	0.9640	0.9964	0.9225	0.9767
5	0.9885	0.9996	0.9646	0.9890
6	0.9964	1.0000	0.9840	0.9948
7	0.9989	1.0000	0.9928	0.9976
8	0.9997	1.0000	0.9968	0.9989
9	1.0000	1.0000	0.9986	0.9995
10	1.0000	1.0000	0.9994	0.9998

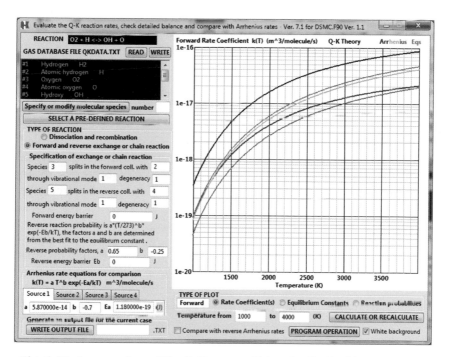

Fig. 6.15 A comparison of the Q-K rate coefficient for $O_2+H\rightarrow OH+O$ with four rates based on commonly used Arrhenius equations.

The DSMC condition, based on the Q-K theory, for the occurrence of the forward reaction is defined in Eqn. (3.45). Then, with the assumption of an equilibrium gas, Eqn. (3.46) provides an analytical prediction of the forward rate equation. The predicted rate is compared with a set of measured rates in the **QKrates** screen image that is shown as Fig. 6.15. The predicted rate is about three times larger than the rate that would be in best agreement with the measured rates. There is a provision for the application of a reduction factor to the rates and an empirical factor of one third could be applied to this reaction. However, empirical factors are not employed in either this or the following sections.

The law of mass action can be written for this reaction as

$$K_{eq} \equiv \frac{k_f}{k_r} = \frac{n_{OH} n_O}{n_{O_2} n_H} = \left(\frac{m_{OH} m_O}{m_{O_2} m_H} \right)^{3/2} \frac{Z_{rot}^{OH} Z_{vib}^{OH} Z_{el}^{OH} Z_{rot}^{O} Z_{vib}^{O} Z_{el}^{O}}{Z_{rot}^{O_2} Z_{vib}^{O_2} Z_{el}^{O_2} Z_{rot}^{H} Z_{vib}^{H} Z_{el}^{H}} \exp\left(-\frac{E_a}{kT} \right). \quad (6.4)$$

The equilibrium constant is plotted from **QKrates** in Fig. 6.16.

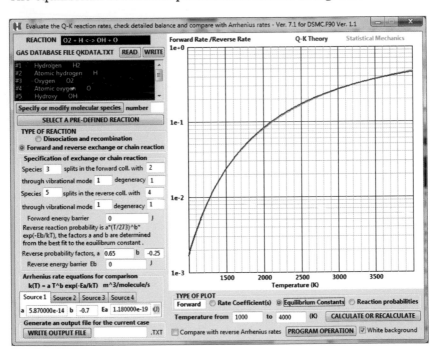

Fig. 6.16 Interactive determination of the reverse reaction probability factors.

Figure 6.16 also shows that the ratio of the forward to reverse reaction rates from the Q-K theory is bought into agreement with that from the law of mass action if the reverse probability factors are $a = 0.65$ and $b = -0.25$. As with the dissociation-recombination case, the agreement is almost exact over the temperature range of interest. Boyd (2007) has shown that it is not possible to exactly satisfy the law of mass action when the both the forward and reverse rates are in the form of the Arrhenius equation. Because the energy barrier for the reverse reaction has been set to the default value of zero, aT^b is in the form of the Arrhenius equation, but it applies to the probability of reaction in a collision rather than to the reaction rate coefficient.

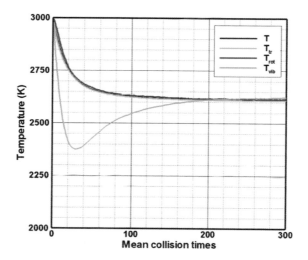

Fig. 6.17 Modal temperatures in the $O_2+H \rightarrow OH+O$ reaction in a mixture of equal parts oxygen and atomic hydrogen initially at a number density of 10^{20} m^{-3} and a temperature of 3,000 K.

The relaxation of the overall temperatures of the energy modes in the test case is shown in Fig. 6.17 and Fig. 6.18 shows the similar relaxation of the species temperatures. The gas comes to equilibrium at a temperature of 2,610 K with 70% oxygen and atomic hydrogen and 30 % hydroxyl and atomic nitrogen. This ratio corresponds to an equilibrium constant of 0.184, while the predicted equilibrium constant from statistical mechanics is 0.182. This result is from the optional output file that is more precise than the plot in Fig. 6.16.

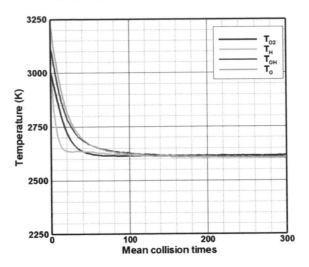

Fig. 6.18 Modal temperatures in the $O_2+H\rightarrow OH+O$ reaction test case.

All the temperatures come to equilibrium to within ±10 K. This is a better result than that for this test case in Bird (2012) where there were continuing differences of 50 K. In addition, the equilibrium temperature in the earlier calculation was 2,420 and the species ratio was 6:4 rather than 7:3. Part of this difference is due to the changes in the Q-K procedures, but most of it is due to the inclusion of the higher electronic levels in the evaluation of the equilibrium constant in the earlier calculation, even though electronic energy was neglected in the DSMC calculation. The current calculation also neglects the electronic energy, but the calculation is consistent in that in that only the ground state degeneracies are taken into account in the evaluation of the equilibrium constant. The electronic energy should ideally be included in the simulation of this reaction because both atomic oxygen and hydroxyl have low lying electronic states. The electronic energy has been excluded from the oxygen-hydrogen model that is to be used in the combustion studies in the following sections because the necessary data does not appear to be available for the triatomic molecules. The effect of electronic energy on hydrogen-oxygen combustion is a subject for future study. These studies should also include thermal radiation. This has been included in some earlier DSMC studies and is discussed in Bird (1994).

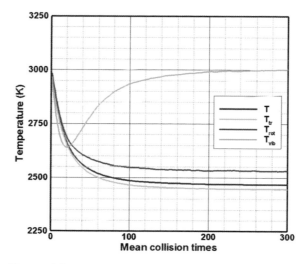

Fig. 6.19 The modal temperatures when the post-reaction vibrational states are set from the Larsen-Borgnakke distributions rather than from the vibrationally-resolved Q-K distributions.

The test case was repeated with the procedures for imposing detailed balance bypassed and, instead, a Larsen-Borgnakke redistribution of the internal energies was made after every reaction. Figure 6.19 shows that the failure to achieve detailed balance in reactions leads serious inter-modal temperature nonequilibrium when the gas has come to a state of chemical equilibrium. A reaction is in chemical equilibrium when there are equal numbers of forward and reverse reactions.

Detailed balance is achieved only because the Q-K theory for gas phase chemical reactions provides analytical expressions for the vibrationally-resolved reaction rates. There do not appear to any alternative analytic expressions in the chemistry literature and the alternative would be to make use of any available numerical data from mainstream chemistry methods that calculate collision trajectories on potential energy surfaces. The quasi classical trajectory, or QCT, method is the best known of these and Bird (2012b) has shown that QCT and Q-K results are in good agreement for this reaction. On the other hand, current QCT results (Wysong et al, 2012) for nitrogen dissociation are not in agreement with the Q-K predictions. However, Bird (2012a) has shown that shock tube measurements of nitrogen dissociation rates support the Q-K rather than the QCT predictions.

6.6 Spontaneous combustion

The test cases in this and the following section are concerned with oxygen-hydrogen reactions. The combustion of an oxygen-hydrogen mixture with water as the product is often written $2H_2 + O_2 \rightarrow 2H_2O$, but the spin states of the commonly occurring forms of these molecules is such that the direct reaction is a forbidden transition. This means that the reaction must take place through a chain reaction that involves additional molecular species. The **OXYGEN_HYDROGEN** gas data subroutine contains seven active molecular species plus argon as an inactive diluent. These have the following code numbers:

#1	H_2	molecular hydrogen
#2	H	atomic hydrogen
#3	O_2	molecular oxygen
#4	O	atomic oxygen
#5	OH	hydroxyl
#6	H_2O	water vapour
#7	HO_2	hydroperoxyl
#8	Ar	argon

The reactions in the gas model include the dissociation and recombination of the five diatomic or triatomic species plus eight pairs of binary exchange or chain reactions. The sixteen binary reactions are given code numbers with those of the forward or endothermic tractions being odd and those of the reverse or exothermic reactions being even. These reactions are:

#1	$H_2 + O_2 \leftrightarrow HO_2 + H$	#2
#3	$O_2 + H \leftrightarrow OH + O$	#4
#5	$H_2 + O \leftrightarrow OH + H$	#6
#7	$H_2O + H \leftrightarrow OH + H_2$	#8
#9	$H_2O + O \leftrightarrow OH + OH$	#10
#11	$OH + OH \leftrightarrow HO_2 + H$	#12
#13	$H_2O + O \leftrightarrow HO_2 + H$	#14
#15	$OH + O_2 \leftrightarrow HO_2 + O$	#16

Reaction #1 is the chain initiating reaction in this set. It leads to two radicals and must always be the first reaction that occurs. Reactions involving hydrogen peroxide provide an alternative means of initiating combustion, but the subsequent combustion process and final state are not sensitive to the manner in which the first radicals are produced. Reactions #3 and # 5 are expected to be the main chain branching reactions and the most important chain terminating reaction that produces a water molecule is #8.

The first test case is a stoichiometric mixture of molecular hydrogen and oxygen at standard number density and a temperature of 1,500 K. This is reported in **DS0D.TXT** as:

```
Data summary for program DSMC
  The n in version number n.m is           1
  The m in version number n.m is           1
  The approximate number of megabytes in the calculation is        100
        8          Oxygen-hydrogen
  The gas properties are:-
    The stream number density is   2.686800000000000E+025
    The stream temperature is    1500.00000000000
    The stream vibrational and electronic temperature is
  1500.00000000000
        The fraction of species        1  is   0.666700000000000
        The fraction of species        2  is   0.000000000000000E+000
        The fraction of species        3  is   0.333300000000000
        The fraction of species        4  is   0.000000000000000E+000
        The fraction of species        5  is   0.000000000000000E+000
        The fraction of species        6  is   0.000000000000000E+000
        The fraction of species        7  is   0.000000000000000E+000
        The fraction of species        8  is   0.000000000000000E+000
```

The temperature history of the gas is shown in Fig. 6.20. Based on the maximum slope thickness of the temperature rise, spontaneous combustion has occurred over a time interval of 20 nanoseconds and after an ignition delay or induction time of 0.17 μs.

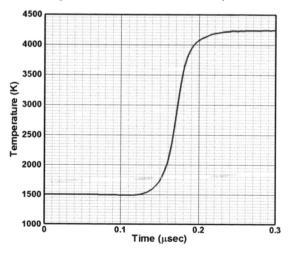

Fig. 6.20 The temperature increase due to spontaneous combustion.

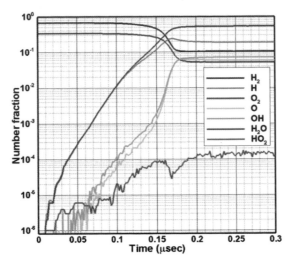

Fig. 6.21 The time history of the number fractions of the molecular species.

The term "ignition delay time" is misleading because, at the level of one molecule in a million, Fig. 6.21 shows that there is almost immediate auto-ignition. The initial number of molecules is one million and the plot in Fig. 6.21 has a resolution slightly better than one in a million. However, the composition is recorded at discrete time intervals and the molecules that appear and disappear within an output interval are not shown. It is worth noting the composition and the number of reactions to have occurred when the number of water molecules has reached 1059. The atomic hydrogen number is then also 1059, with 3 molecules of atomic oxygen, 16 hydroxyl molecules and 5 molecules of hydroperoxyl. There had been 8 #1, 4 #2 reactions and one recombination to hydroperoxyl. The dominant reactions were, as expected, #3 with 539, #5 with 536, and # 8 with 1059.

The temperature equivalent E_a/k of the activation energy of the initiating reaction #1 is 27,700 K and the Q-K condition for the reaction to occur is applied only when the sum of the translational collision energy and the vibrational energy in the hydrogen molecule exceeds this value. Figure 2.2 plots, from Eqn. (2.35), the fraction of collisions in an equilibrium gas that has a relative translational energy above a specified value of the ratio $E_a/(kT)$. For reaction #1 at 1,500 K, this ratio is 18.47, but the vibrational energy supplies an additional kT so that the effective ratio is about 17.5. The fraction of collisions with energy above the activation energy is then about two or three times 10^{-7} and about one tenth of these will pass the Q-K reaction selection test of Eqn. (3.45). The mean collision time for H2-O2 collisions at 1,500K is about 1 nanosecond so that a reaction can be expected after twenty or thirty nanoseconds. The initiating reaction in Fig. 6.21 has occurred slightly earlier than expected. The result from a similar, but statistically independent, calculation is shown in Fig. 6.22 and, in this case, the reaction has occurred at about the expected time. Apart from this variable delay in the ignition, the combustion process is repeatable.

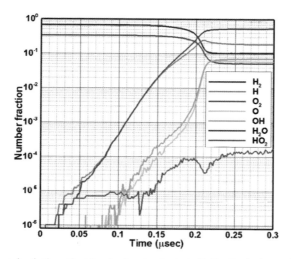

Fig. 6.22 A calculation that is similar to, but statistically independent of, the calculation in Fig. 6.21 .

The volume in a homogeneous gas calculation is unspecified and the number density can be set to any desired value. However, for combustion calculations, there is an unphysical delay in the initiation of the reaction unless the total number of molecules is of the order of the inverse of the reaction probability in a collision. The number in this test case is marginal, but it turns out that the overall duration of the reaction process is remarkably insensitive to the number of simulated molecules. While more molecules lead to earlier ignition, it occurs at a lower concentration of radicals and time is required to reach the higher concentration of radicals that occurs in an ignition with fewer molecules.

The subsequent rate of increase in the number density of water vapour is exponential. In both calculations, the water vapour number fraction increased by a factor of ten in approximately 0.03 microseconds. Because reaction #8 dominates, the traces for water vapour and atomic hydrogen coincide for most of the reaction process.

The initial temperature of 1,500 K is well above the observed explosion limit that is around 850 K (Glassman, 1996). At this temperature, the $E_a/(kT)$ ratio increases from 18.5 to about 32.5 and Fig. 2.2 indicates that the required number of simulated molecules would have to be increased by a factor of 10^6. Moreover, the required length of the calculation (in terms of mean collision times) would increase by a similar factor and calculations near the explosion limit are clearly inaccessible to DSMC. On the other hand, the ratio decreases to 13.85 at, for example, 2,000 K and calculations are readily made at higher temperatures. The test calculation has therefore been repeated for temperatures ranging from 1,300 K to 4,000 K and simulated molecule numbers ranging from 100,000 to 10,0000,000. The ignition delay time is set to the time at which the water vapour concentration reaches 30% by number. Delay times have traditionally been plotted against one thousand divided by the pre–combustion gas temperature in degrees K. The measurements have then been found to fall on a straight line.

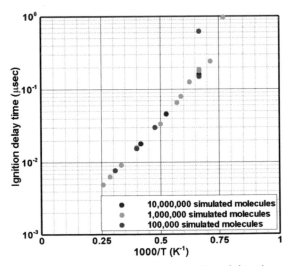

Fig. 6.23 The traditional plot for ignition delay times.

Figure 6.23 shows that, while there is scatter at the lower temperatures, the delay times from the DSMC simulation also fall on a straight line. For temperatures above 2,000 K, the results are repeatable and independent of the number of simulated molecules. The physical conclusion from this figure is that, no matter how high the pre-combustion temperature, the ignition delay time would be expected to not fall below one nanosecond. It is more interesting to determine whether any prediction of the explosion limit can be made by extrapolation of the results from DSMC calculations that can be readily made on a PC. The extrapolation should exclude the data points that have been seriously affected by the unphysical delay that occurs because of an insufficient number of simulated molecules. The two highest points in Fig. 6.24 would be excluded on this basis, and while the computational limit on the temperature is too high for a confident extrapolation, the results are consistent with the observed explosion limit.

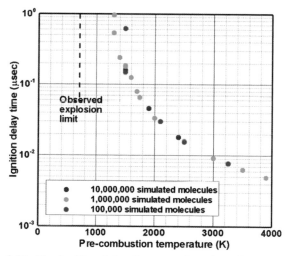

Fig. 6.24 The ignition delay time as a function of temperature.

The calculations employ a stoichiometric mixture and the degree of completeness of the reaction should be compared with existing data. The available data is for the flame temperatures that result from combustion at lower pre-combustion temperatures. There is a greater degree of dissociation at the higher temperatures and the decrease in the equilibrium number fraction of water vapour is shown in Fig. 6.24.

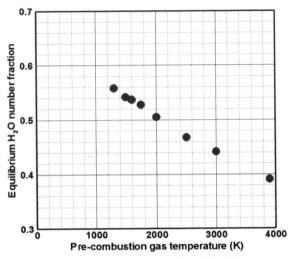

Fig. 6.25 Effect of temperature on the yield of water vapour.

The equilibrium number fraction of H_2O in the case shown in Figs. 6.21 and 6.22 is 0.54. An initial calculation for this case was made with all the energy barriers set to the default value of zero and the equilibrium H_2O fraction was only 0.4. This was thought to be too low and an inspection of the equilibrium reaction numbers showed that the H_2O depleting reaction #7 occurred almost as frequently as reaction #8. An energy barrier of 1×10^{-19} J was therefore applied to reaction #7 as the only change to the default data. The resulting number of reactions between 0.3 and 0.4 μs in a continuation of the Fig. 6.22 case was then reported in an output file as:

```
Total simulated molecules =    873074

Species          1  total =       90817
Species          2  total =      160380
Species          3  total =       44603
Species          4  total =       50486
Species          5  total =       62082
Species          6  total =      464594
Species          7  total =         112
Species          8  total =           0

SP        1 DISSOCS  37715.0000000000    RECOMBS  44319.0000000000
SP        3 DISSOCS   2456.00000000000    RECOMBS    828.000000000000
SP        5 DISSOCS  19543.0000000000    RECOMBS  49340.0000000000
SP        6 DISSOCS  40181.0000000000    RECOMBS  15744.0000000000
SP        7 DISSOCS   9873.00000000000    RECOMBS     19.0000000000000
```

```
EX,C reaction      1   number    14226.0000000000
EX,C reaction      2   number    99572.0000000000
EX,C reaction      3   number    1457862.00000000
EX,C reaction      4   number    1358339.00000000
EX,C reaction      5   number    6497492.00000000
EX,C reaction      6   number    9899119.00000000
EX,C reaction      7   number    2080476.00000000
EX,C reaction      8   number    5574216.00000000
EX,C reaction      9   number    4129120.00000000
EX,C reaction     10   number    733856.000000000
EX,C reaction     11   number    27488.0000000000
EX,C reaction     12   number    0.000000000000000E+000
EX,C reaction     13   number    73558.0000000000
EX,C reaction     14   number    0.000000000000000E+000
EX,C reaction     15   number    4867.0000000000
EX,C reaction     16   number    10742.0000000000
```

With the energy barrier on reaction #7, the number of #8 reactions is almost three times more numerous than #7 reactions. This energy barrier is almost as large as the heat of reaction and is larger than that the barrier that generally appears in the Arrhenius equations. However, it has to be emphasised that, as well as modelling observed barriers, the energy barriers are employed to compensate for any clear deficiencies in the Q-K rates.

The available data indicates that an energy barrier should also be applied to reaction #8, but the objective of these calculations is to demonstrate DSMC capabilities rather than to develop the most realistic model. However, some sensitivity studies were made and it was found that, while the latter stages of the reaction directly reflect the rate of reaction #8, it has little effect during the exponential growth stage. Instead, the slope of the exponential H_2O trace is set largely by the rates of the chain branching reactions #3 and #5.

6.7 Forced ignition

As discussed in the preceding section, it has not been possible to attain spontaneous combustion in DSMC simulations when the pre-combustion temperature is less than 1,300 K. This is due to the fact that the probability of finding a hydrogen molecule in about the fifth vibrational energy level decreases exponentially with temperature. It should therefore be possible to extend the lower temperature limit by "spiking" the initial equilibrium gas with a small number of molecules with enhanced vibrational energy. Special coding was therefore added at the end of the **SET_INITIAL_STATE_1D** subroutine to reset to five the vibrational level of a specified percentage of the initial molecules.

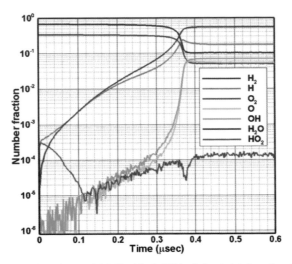

Fig. 6.26 Ignition from 1200 K with 0.05% of the initial molecules set to vibrational level 5.

Figure 6.26 shows the time history of the composition when only 500 of the initial 1,000,000 simulated molecules have elevated vibrational levels. The overall temperature increase due to the added energy was just over five degrees Kelvin. Reaction #1 occurred in the initial collisions of 300 molecules and chain branching reactions led the water vapour concentration to quickly catch up with the atomic nitrogen concentration. The hydroperoxyl concentration dropped to that of atomic oxygen and hydroxyl. While the initiation processes are qualitatively different, the final stage of the combustion process is similar to that for spontaneous combustion. Because the reaction rates are the same, the species concentrations in the equilibrium post-combustion gas are affected only by the differences in the equilibrium temperature.

A similar calculation was made for an initial temperature of 1,000 K and the results are plotted in Fig. 6.27. The major difference is that the time to reach a water vapour number fraction of 0.3 increases from 0.36 µs to 1.175 µs. As with spontaneous combustion, this can be regarded as an ignition delay time. The initial temperature was then reduced by a further 200 K to 800 K. As shown in Fig. 6.27, this resulted in a dramatic increase in the ignition delay time to 29.0 µs. This calculation involved the computation of 3.5×10^{11} collisions and the computation time on a single CPU was of the order of a week.

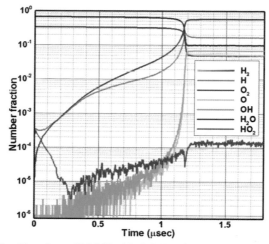

Fig. 6.27 Ignition from 1000 K with 0.05% of the initial molecules set to vibrational level 5.

After a delay of 29 μs, the combustion process appears to be an almost instantaneous event. The temperature profile is shown in Fig 6.29 and there is a gradual build-up of temperature during the delay time. The time from a temperature of 1,000 K to combustion is 1.5 μs and is comparable with that in Fig 6.27.

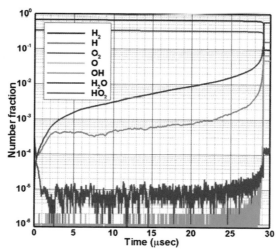

Fig. 6.28 Ignition from 800 K with 0.05% of the initial molecules set to vibrational level 5.

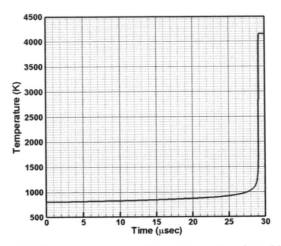

Fig. 6.29 The temperature profile in the ignition from 800 K.

In order to determine whether the delay time in forced ignition is a function of ignition energy as well as pre-combustion temperature, the case of Fig 6.28 was repeated with ten times as many initially excited molecules. The excitation energy is then equivalent to a temperature increase of 53 K and Fig 6.29 shows a fourfold reduction in the delay time.

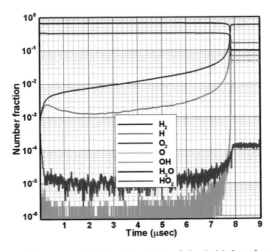

Fig. 6.29 Ignition from 800 K with 0.5% of the initial molecules set to vibrational level 5.

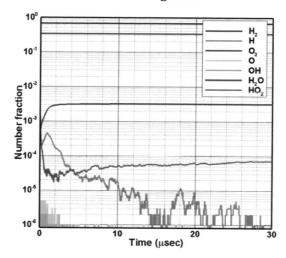

Fig. 6.30 Failed ignition from 300 K with 0.5% of the initial molecules set to vibrational level 5.

The higher level of ignition energy was insufficient to attain ignition from an initial temperature of 300K. Figure 6.30 shows that the initial concentrations of water vapour was similar to that in the 800 K case but, at about 1 μs, there was a sharp reversal of the rate of change of hydrogen atoms. The number of water vapour molecules and, therefore, the temperature of the gas quickly levelled off and there was no combustion. In a similar calculation with only 0.05% initially excited molecules, the number of hydrogen atoms declined to zero and all reactions then ceased. However, in this case, a small number of reactions were present until the calculation was terminated at 30 μs. The active reactions were the initiation reaction #1, its reverse #2 and the recombination of hydroperoxyl. The initial reactions were sufficient to raise the temperature to 380 K, but the activation energy for reaction #1 is so high that it occurred many orders of magnitude more frequently than expected. The probable reason for this is that the hydroperoxyl recombinations led to a few molecules with exceptionally high energies.

The continued presence of hydroperoxyl is because the unimolecular dissociation of this unstable molecule was not taken into account. Given the availability of data on the probability of this reaction, the program logic could readily be extended to include it. Other than the setting of an energy barrier to reaction #7, the calculations have employed the default data. The objective has been to demonstrate the capabilities of DSMC in the context of combustion studies and, for detailed comparisons with experiment, far more attention should be given to the optimization of the data. While the Q-K theory permits computations to be made in the absence of rate data based on experiment, the adjustment of the energy barriers permits the rates to be bought into agreement with any rates based on reliable experiments. This is facilitated by the **QKrates** program.

Note that, while there can be continuum studies of spontaneous combustion based on the rate equations, forced ignition through excited vibration appears to require a particle approach. Feasible extensions of these calculations would be directly relevant to work on laser ignition.

References

Bird, G. A. (1994). *Molecular Gas Dynamics and the Direct Simulation of Gas Flows,* Clarendon Press, Oxford.

Bird, G. A. (2009) A Comparison of Collision Energy-based and Temperature based Procedures in DSMC, in *Rarefied Gas Dynamics* (Ed. T. Abe), American Institute of Physics Conference Proceedings 1084, 245-250.

Bird, G. A. (2011a) Chemical Reactions in DSMC, in *Rarefied Gas Dynamics* (Ed. D.A. Levin, I.J. Wysong and A.L. Garcia), American Institute of Physics Conference Proceedings 1333, 1195-1202.

Bird, G. A. (2011b) The Q-K model for gas-phase chemical reaction rates, *Phys. Fluids* **23**, 106101

Bird, G. A. (2012a) DSMC Simulations of Shock Tube Experiments for the Dissociation Rate of Nitrogen, *28th Symposium on Rarefied Gas Dynamics*, AIP Conference Proceedings 1501, 595.

Bird, G. A. (2012b) Setting the post-reaction internal energies in direct simulation Monte Carlo simulations. *Phys Fluids* **24**, .

Boyd, I. D. (2007) Modeling Backward Chemical Rate Processes in the Direct Simulation Monte Carlo Method. *Phys. Fluids.* **19**, 126103 (2007).

Glassman, I. (1996). *Combustion,* Third edition, Academic Press, San Diego.

Haas, B. L., Hash, D., Bird, G. A., Lumpkin, F. E., and Hassan, H. (1994). Rates of thermal relaxation in Direct Simulation Monte Carlo Methods, *Phys. Fluids* **6**, 2191-2201.

Wysong, I., Gimelshein, S., Gimelshein, N., McKeon, W. and Esposito, F. Reaction cross sections for two direct simulation Monte Carlo models: Accuracy and sensitivity analysis, *Phys. Fluids* **24**, 042002 (2012).

Vincenti, W. G. and Kruger, C. H. (1965). *Introduction to Physical Gas Dynamics,* Wiley, New York.

7

ONE-DIMENSIONAL FLOW APPLICATIONS

7.1 Description of the DS1D.DAT data file

The general comments at the start of §6.1 apply to all data files.

NVER   the n in the **DSMC.F90** version $n.m$ for which the file was generated.

MVER   the m in the **DSMC.F90** version $n.m$ for which the file was generated.

IMEG   approximate number of megabytes to be used by the program. (The initial number of simulated molecules will be 10,000 times **IMEG**)

IGAS   the code number of the gas data subroutine that is to be employed in the calculation. The current codes are:

1 d = 4e-10 m hard sphere [**MSP=1, MMVM=0**]
2 argon [**MSP=1, MMVM=0**]
3 ideal nitrogen [**MSP=1, MMVM=0**]
4 real nitrogen [**MSP=1, MMVM=1**]
5 for ideal air [**MSP=2, MMVM=0**]
6 real air appropriate to re-entry speeds [**MSP=5, MMVM=1**]
7 helium-xenon mixture [**MSP=2, MMVM=0**]
8 combustible hydrogen-oxygen mixture [**MSP=8, MMVM=3**]

FND(1)   the stream or reference number density.

FTMP(1)   the stream or reference temperature.

If **MMVM=0**,   **FVTMP(1)** the vibrational temperature.

VFX(1)   the velocity component in the x direction.

VFY(1)   the velocity component in the y direction.

If **MSP>1** *in a loop* **n** *over* **MSP**

  FSP(1,n)   the fraction of this species in the stream.

end of loop over molecular species.

IFX   0 for a plane flow.

           1 for a cylindrical flow with center at x=0.

           2 for a spherical flow with center at x=0.

XB(1)   minimum x coordinate of the flow (must be ≥ 0 if **IFX>0**).

ITB(1)   0   the minimum coordinate is a stream boundary.

               1   it is a plane of symmetry.

               2   it is a solid surface.

               3   it is an interface with a vacuum.

               4   it is an axis or center.

If **ITB(1)=2** *then*

    TSURF(1)   the surface temperature

    FSPEC(1)   the fraction of specular reflection

    VSURF(1) the velocity component in the y direction

end if

XB(2)   maximum x coordinate of the flow (must be >0 if **IFX>0**).

ITB(2)   0   the maximum coordinate is a stream boundary.

               1   it is a plane of symmetry.

               2   it is a solid surface.

               3   it is an interface with a vacuum.

               4   stream boundary with constant total molecules.

                  (This is the secondary stream if one is present)

If **ITB(2)=2** *then*

    TSURF(2)   the surface temperature

    FSPEC(2)   the fraction of specular reflection

    VSURF(2)   the velocity component in the y direction

end if

If **IFX>0** *then*

    IWF   0   if there are no radial weighting factors.

             1   with weighting factors **1+(WFM -1)**$\{x/$ **XB(2)** $\}^{\textbf{IFX}}$.

    If **IWF=1   WFM**   the maximum weighting factor.

end if

IGS   0   if the initial state of the flowfield is a vacuum.

           1   if it is the stream and any secondary stream.

ISECS   0   if there is no secondary stream.

             1   if there is a secondary stream (**IWF** must be 0).

If **ISECS=1** *then*

(The initial boundary between the main and secondary streams is at x=0 in a plane flow [**IFX=0**])

If **IFX>0   XS** initial boundary between the main and sec. streams.

FND(2) the number density of the secondary stream.

FTMP(2)   the temperature of the secondary stream.

If **MMVM>0,   FVTMP(2)** the stream vibrational temperature.

VFX(2) the velocity component in the x direction.

VFY(2) the velocity component in the y direction.

If **MSP>1** *in a loop* **n** *over* **MSP**

　　FSP(2,n)　　the fraction of this species in the sec. stream.

end of loop over molecular species.

end if

If **IFX=0** *and* **ITB(1)=0** *then*

　　IREM　　**0**　　if there is no molecule removal.

　　　　　　　1　　for removal to keep the total number constant.

　　　　　　　2　　for removal as a restart option

　　If **IREM=1**　　**XREM**　　x coordinate at which removal commences

end if

If **ITB(2)=1** *then*

　　IVB　　**0**　　if the outer boundary is stationary.

　　　　　　1　　if the outer boundary has a constant speed.

　　If **IVB=1**　　**VWALL**　　the speed of the outer boundary

end if

MOLSC　　the desired number of molecules in a sampling cell

7.2 Statistical scatter and random walks

The most useful test case for any DSMC program is a steady uniform flow because one can have complete confidence in the correct set of results. Microscopic fluctuations are present in real gases and the term "statistical scatter" refers to the enhanced fluctuations that are present in DSMC calculations that model the real molecules by a relatively very small number of simulated molecules. Random walks are an unphysical effect that results from the presence of repetitive numerical procedures that, while they are correct on the average, involve a departure from the average at each application. The two effects have been discussed in Sections 1.1, 1.2 and 4.1.

The main test case is a uniform hard sphere gas with cylindrical symmetry that extends from the axis to a solid diffusely reflecting surface at the same temperature as the gas. The data for this case was reported by **DS1D.TXT** as:

```
Data summary for program DSMC
 The n in version number n.m is          1
 The m in version number n.m is          1
 The approximate number of megabytes for the calculation is      10
          1         Hard sphere gas
 The gas properties are:-
```

```
The stream number density is  7.033750000000000E+019
The stream temperature is   300.000000000000
The stream velocity in the x direction is  0.000000000000000E+000
The stream velocity in the y direction is  0.000000000000000E+000
Cylindrical flow
The minimum x coordinate is  0.000000000000000E+000
The minimum x coordinate is an axis or center
The maximum x coordinate is   1.00000000000000
The maximum x coordinate is a solid surface
The maximum x boundary is a surface with the following properties
    The temperature of the surface is   300.000000000000
    The fraction of specular reflection at this surface is  0.000000E+000
    The velocity in the y direction of this surface is  0.0000000000E+000
There are radial weighting factors
The maximum value of the weighting factor is    1000.00000000000
The flowfield is initially the stream(s) or reference gas
There is no secondary stream initially at x > 0
The desired number of molecules in a sampling cell is       1000
```

The ten megabytes leads to an initial 100,000 simulated molecules and the number of molecules in a sampling cell has been set to 1000. This means that there are 100 sampling cells of uniform width and the area of the cell at the axis is only 1/199 that of the outermost cell. The molecule number in a uniform gas is proportional to the cell area so that the statistical scatter at the axis is more than ten times that at the periphery. To avoid the enhanced statistical scatter near the axis, weighting factors have been introduced such that the number of real molecules that are represented by each simulated molecule is directly proportional to the radius. To avoid a zero weighting factor at the axis, the DSMC.F90 program employs weighting factors defined by

$$W = 1 + \left(W_{max} - 1\right)\left(r / r_{max}\right)^{\varepsilon}, \qquad (7.1)$$

where ε is one for cylindrical flows and two for spherical flows. The behavior of weighting factors in axially-symmetric flows can be expected to be similar to that in the one-dimensional unsteady cylindrical flows that are considered in this section.

The maximum weighting factor W_{max} is a data item and Fig. 7.1 shows the extent to which this can be employed to equalize the sample size for the flow properties. The simulated molecule distribution is near uniform for sufficiently high values of W_{max}, but there is a large change in weighting factor across the cell nearest to the axis and this can be expected to lead to unpredictable anomalies.

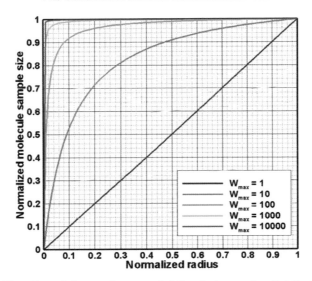

Fig. 7.1 The effect of the maximum weighting factor on the distribution of the normalized sample size of simulated molecules across a cylindrical flow.

In the presence of weighting factors, the number of real molecules that are represented by each simulated molecule varies with radius. Should a molecule move further from the axis and the ratio of the new to the old weighting factor is, say, 1.1 there must be a ten percent chance of it being removed. Conversely, should it move closer to the axis and the ratio is 0.9, there must be a ten percent chance of it being duplicated. Because of the duplication and removal of molecules, the total number of molecules varies even when, as in this case, there are no open boundaries. This variation can be expected to have the characteristics of a random walk with the possibility of large excursions from the nominal number. There is an optional output file for the total number of molecules as a function of time and the results from this are plotted in Fig. 7.2. The time is normalized by the flow radius divided by the most probable molecular speed and, for this problem, the "flow transit time" is equivalent to 17.5 mean collision times. While the flow is of unusually long duration, the magnitude of the random walk excursions in the total molecule number is unacceptable. Weighting factors should be used only for flows of relatively short duration or for flows with high speed stream boundaries such that the molecules are constantly refreshed.

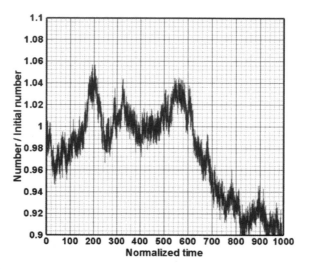

Fig. 7.2 The variation of the total molecule number with time.

A similar calculation was made with a maximum weighting factor of 10 rather than 1000. The excursion in the total molecule number was even larger than that in Fig. 7.2.

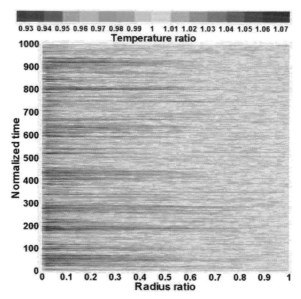

Fig. 7.3 An *x-t* diagram of the temperature fluctuations.

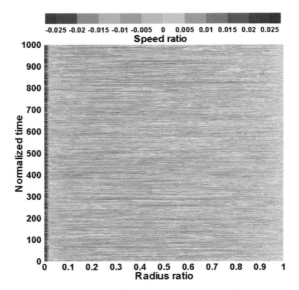

Fig. 7.4 An *x-t* diagram of the speed ratio fluctuations.

The random walk in the total molecule number density would largely obscure the statistical scatter of the number density, but the fluctuations in the other flow properties are worthy of investigation. Figures 7.3 and 7.4 present distance-time representations of the scatter in temperature and flow speed, respectively. These are based on the file **DS1xt.DAT** that is optionally output from **DSMC.F90** for continuing unsteady flows. The flow speed is normalized by the most probable molecular speed in the initial gas and the temperature by the initial gas temperature. The sample size at all data points is approximately 7,500 and the expected standard deviation of the statistical scatter is just over one percent. There are 100 cell of uniform with in the radial direction, but about 2,000 output intervals at uniform time steps. This accounts for the almost horizontal appearance of the fluctuation structures.

The temperature scatter in the outer 40% of the flowfield is generally in line with expectations, but is systematically higher toward the axis. The innermost sampling cell shows some additional effects, but is not seriously affected. On the other hand, the speed ratio in the innermost cell is systematically enhanced by about two standard deviations. Outside that cell, the scatter is in line with expectations and is unaffected by radius.

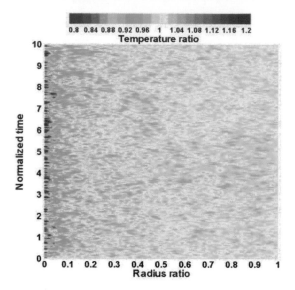

Fig. 7.5 The temperature fluctuations in the absence of weighting factors.

A similar calculation was made without the application of weighting factors and the statistical scatter in the temperature and speed ratio is shown in Figs. 7.5 and 7.6.

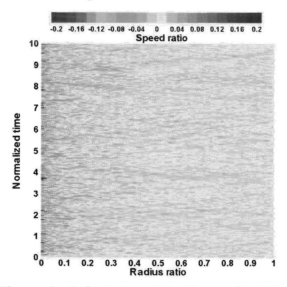

Fig. 7.6 The speed ratio fluctuations in the absence of weighting factors.

The approximate sample size for the calculation of the flow properties ranged from 50,000 in the outermost sampling cells down to about 250 in the innermost cell. This jumps to 750 in the second cell and is more than 2500 at the sixth cell. The statistical scatter is proportional to the square root of the sample size so that the unacceptable level of scatter is confined to the first few cells.

A sound wave traverses the flow in a normalized time of 0.913 and a remarkable feature of the velocity fluctuation contours is that there appears to be structures within the fluctuations that move in either direction with the speed of sound.

7.3 Shock wave structure in a simple gas

The normal shock wave has been the subject (Bird, 1994) of some of the most useful validations of DSMC against experiment. These comparisons were made when there were similar error bars of one or two percent for both DSMC and experiment. DSMC calculations for the structure of normal shock waves are now readily made with an accuracy of ±0.1% and the **DSMC.F90** program has been used to make definitive calculations for weak, moderate and strong shock waves in argon. It should be kept in mind that the uncertainties in the transport properties of argon are of the order of 1% rather than 0.1%.

Because the upstream and downstream flow conditions are known from the Rankine-Hugoniot relations, the simplest strategy is to specify dual steam initial conditions with a discontinuous change at the origin. The "eventual steady flow" run-time option is selected and the discontinuity evolves into a shock wave. At some time after the completion of the unsteady phase, the program is stopped and restarted with a new sample. The standard stream entry procedures would, on the average, maintain the molecule number at its initial value. However, this number would be subject to a random walk at every time step and any significant departure from the average would result in movement of the shock wave. A special boundary type option has therefore been added especially for shock wave problems. This adjusts the molecule entry number at the downstream boundary to values that exactly maintain the overall number. However, the entry number cannot be negative and there will occasionally be momentary increase in the molecule number above its nominal value.

Profiles have been calculated for weak, moderate and strong shock waves. The data for the moderate shock Mach number 2 case is:

```
Data summary for program DSMC
 The n in version number n.m is            1
 The m in version number n.m is            1
 The approximate number of megabytes for the calculation is        10
          2       Argon
 The gas properties are:-
     The stream number density is  1.294384000000000E+018
     The stream temperature is   273.000000000000
     The stream velocity in the x direction is   615.632100000000
     The stream velocity in the y direction is  0.000000000000000E+000
 Plane flow
 The minimum x coordinate is  -40.0000000000000
 The minimum x coordinate is a stream boundary
 The maximum x coordinate is   20.0000000000000
 The maximum x coordinate is a stream boundary with a fixed number of
 simulated  molecules
 The flowfield is initially the stream(s) or reference gas
 There is a secondary stream applied initially at x = 0 (XB(2) must be > 0)
 The secondary stream (at x>0 or X>XS) properties are:-
     The stream number density is  2.958587000000000E+018
     The stream temperature is   567.329300000000
     The stream velocity in the x direction is   269.339500000000
     The stream velocity in the y direction is  0.000000000000000E+000
 There is no molecule removal
 The desired number of molecules in a sampling cell is        250
```

The VHS molecular model has been employed and the non-dimensional number density, temperature, parallel temperature component and normal temperature component are shown in Fig. 7.7.

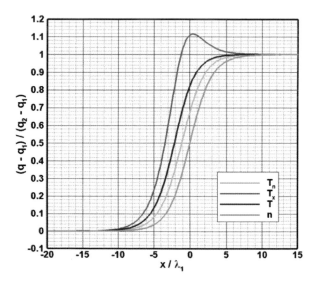

Fig. 7.7 The structure of a shock of Mach number 2 in argon.

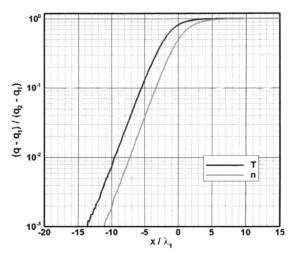

Fig. 7.8 A log plot of the normalized properties.

The upstream extent of the shock appears to be about $x/\lambda_1 = -12$ in Fig. 7.7, but the log plot in Fig. 7.8 indicates that the shock extends indefinitely in the upstream direction. The profiles become very nearly straight lines in this plot so that the normalized flow quantities decline exponentially with upstream distance. The velocity distribution function of a one-dimensional flow is axially-symmetric and, well upstream of the shock wave, it is a spherically symmetric equilibrium distribution relative to the upstream flow on which is superimposed a very sharp spike in the negative x direction. The adequate representation of this distribution represents an almost insuperable difficulty for numerical solutions of the Boltzmann equation for moderate and strong shock waves.

The density profile is very nearly symmetric about the origin, but the temperature profile leads the density profile by about two upstream mean free paths. The degree of translational non-equilibrium in the flow is best represented by the ratio of the temperature based on the velocity components in the flow direction to that based on the components normal to the flow direction. The parallel temperature overshoots the downstream temperature by 11% and, at the origin, it is 70% higher than the normal temperature. This is beyond the limits of the Chapman-Enskog theory and the Navier-Stokes profile is in significant error for a shock Mach number of two.

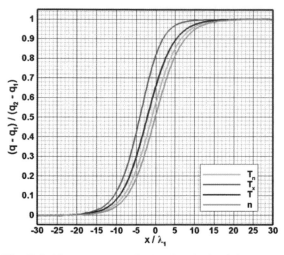

Fig. 7.9 The structure of a Mach 1.3 shock in argon.

The other test cases are for shock Mach numbers of 1.3 and 8. These are the near cube root and cube of the Mach 2 shock that has a temperature ratio of 2.07. The weak shock wave has a temperature ratio of 1.29 and is well within the range of validity of the Navier-Stokes equations. The temperature ratio across the very strong Mach 8 shock is 20.87.

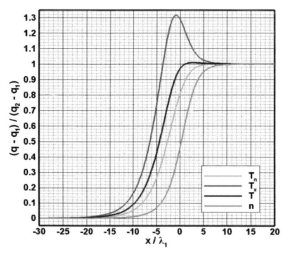

Fig. 7.10 The structure of a Mach 8 shock in argon.

The very strong shock wave is qualitatively different from the weaker waves in that the density profile is asymmetric and there is an overshoot, albeit only 0.8%, in the normalized temperature. Also, while the separation between the temperature and number density profiles is close to two upstream mean free paths in the lower strength shocks, it increases to nine upstream mean free paths in the Mach 8 shock. The overshoot in the normalized parallel temperature is below 0.1% for Mach 1.3, 12% for the Mach 2 shock and 31.5 for the Mach 8 shock.

The ratio of the parallel to the normal temperature increases with the shock Mach number and the question arises as to whether the ratio of these temperatures can be used to predict whether or not the Navier-Stokes equations are valid. Note that it is the ratio of actual, rather than the normalized temperatures, that is being considered. This ratio will fluctuate about unity in the uniform flow regions and it is desirable to subtract one from the ratio. In order to permit the use of a log scale, the absolute value is employed. This suggested non-equilibrium parameter is plotted in Fig. 7.11 for the three cases.

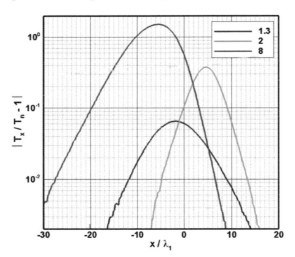

Fig. 7.11 Parameter based on the parallel to normal temperature ratio.

The local Knudsen numbers based on the scale lengths of the gradients in the macroscopic flow properties appear in Eqn. (1.28) for the Chapman-Enskog distribution and provide the most reliable guide to the validity of the Navier-Stokes equations.

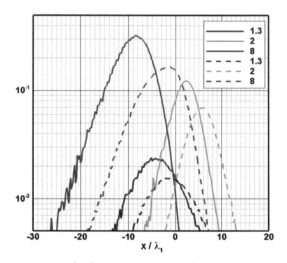

Fig. 7.12 $s\dfrac{\lambda_1}{u}\left|\dfrac{du}{dx}\right|$ (full lines) and $\dfrac{\lambda_1}{T}\left|\dfrac{dT}{dx}\right|$ (dashed lines).

The local Knudsen number based on the velocity scale length is multiplied by the local speed ratio in Eqn. (1.28). The distributions through the shock waves of this product and the local Knudsen number based on the temperature scale length are shown in Fig. 7.12. The other terms in Eqn. (1.28) may be regarded as being of order unity and the condition for the validity of the Chapman-Enskog theory, and therefore of the Navier-Stokes equations, is that the sum of these parameters should be small in comparison with unity. The sum is 0.04 for $M_S = 1.3$, 0.19 for $M_S = 2$ and 0.46 for $M_S = 8$. The corresponding values of the parameter $\left|T_x/T_n - 1\right|$ are 0.065, 0.37 and 1.51. The Navier-Stokes equations are generally considered to be valid for shock Mach numbers up to about 1.8.

The evaluation of gradients is generally avoided in DSMC procedures because very large samples are needed to obtain reasonably smooth curves for the gradients of the macroscopic properties. The ratio of the parallel to the normal temperature is more readily calculated and these calculations suggest that the criterion for the validity of the Navier-Stokes equations is

$$\left|\frac{T_x}{T_n} - 1\right| < 0.1. \tag{7.2}$$

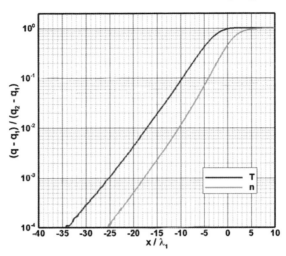

Fig. 7.13 The extent of the upstream influence of the Mach 8 shock wave.

The larger the disturbance, the easier it is to obtain an accurate DSMC solution. The shock Mach number 8 case was continued until the sample size exceeded 10^9, so that the standard deviation in the scatter was less than 0.0001. Figure 7.13 shows that the normalized temperature and density reach this value at x/λ_1 35 and 25, respectively. Unlike the Mach 2 case in Fig. 7.8, there is a slight curvature in the log plot of the density and temperature and, the separation between the curves. The figure is based on a time average from ten to thirty seconds and it was considered that ten seconds may have been too short a time for the far upstream flow to have become steady. The calculation was continued and a new sample was taken from 30 to 35 milliseconds. There was no change in the profiles and the features are evidently valid. They can therefore be extrapolated further upstream and the temperature can be expected to increase by one part in a million at just over 50 mean free paths upstream of the centre of the wave.

The far upstream disturbance occurs in a region where the both the local Knudsen numbers and the parameter in Eqn. (7.2) are very small. However, as discussed earlier, it is a consequence of an extreme non-equilibrium distribution function. Therefore, while the Navier-Stokes equation become invalid when the non-equilibrium parameters become too large, they are not necessarily valid when these parameters are very small.

7.4 Viscosity, heat transfer and diffusion

The diameters of the VHS models have been chosen to match the viscosity and heat transfer coefficients of each molecular species. The VSS model is designed to also match the diffusion coefficients. The **DSMC.F90** program can be used to set up test cases for the verification of the procedures that are employed to set the model parameters.

The vast majority of the formulations of the Navier-Stokes equations include only the three basic transport properties of viscosity, heat transfer and diffusion. Moreover, in the case of diffusion, only **forced diffusion** in gas mixtures is recognized. Chapman and Cowling (1970) write the **relative diffusion velocity** in a binary gas mixture as

$$\boldsymbol{C_1} - \boldsymbol{C_2} = \frac{-n^2}{n_1 n_2} D_{12} \left\{ \nabla \frac{n_1}{n} + \frac{n_1 n_2 (m_2 - m_1)}{n \rho} \nabla (\ln p) \right.$$
$$\left. + k_T \nabla (\ln T) - \frac{\rho_1 \rho_2}{p \rho} (\boldsymbol{F_1} - \boldsymbol{F_2}) \right\}. \tag{7.3}$$

The first of the four terms represents the forced diffusion that acts to reduce a concentration gradient. The last of the four terms represent the species separation that is caused by an external force such as that in the upper atmosphere due to gravitation. The second and third terms describe **pressure diffusion** and **thermal diffusion** that act to produce species separation whenever there are pressure and temperature gradients in the flow. These effects are very rarely included in the continuum models. Thermal diffusion is due to differences in molecular size and was not discovered until it had been predicted by the Chapman-Enskog theory. The diffusion thermo-effect is the inverse of thermal diffusion and is a heat flux in a gas at a uniform temperature when there is a gradient in concentration. These effects are not exclusive and, should it be possible to model the unexpectedly large upstream extent of shock waves by transport terms, they would probably involve the **self-diffusion coefficient**.

The lesser-known transport properties should not be confused with effects such as **thermal creep** that involves finite flow velocities in the vicinity of surfaces with temperature gradients. This is a rarefied flow phenomenon that arises when molecules are able to travel distances comparable to the scale lengths of the surface gradients.

The DSMC method models all the above effects.

Couette flow is the obvious test case for the verification of the viscosity coefficient.   This is the flow that develops between a stationary flat plate and a parallel moving plate.   Consider the Knudsen number 0.005 case in which the stationary plate is in the $x = 0$ plane and is separated by 200 mean free paths from the upper plate that moves in the y direction with a speed ratio of 0.5. As with the shock wave calculation, the VHS model of argon is employed. The gas is initially stationary at 273 K and the number density of 1.2944×10^{18} is such that the mean free path is unity. The surfaces of the plates are diffusely reflecting at a temperature of 273 K and the upper wall starts in moving in the y direction with speed $U_P = 168.595$ m/s at zero time. The data file is reported as:

```
Data summary for program DSMC
  The n in version number n.m is              1
  The m in version number n.m is              1
  The approximate number of megabytes for the calculation is         10
          2   Argon
  The gas properties are:-
      The stream number density is  1.294384000000000E+018
      The stream temperature is   273.000000000000
      The stream velocity in the x direction is  0.000000000000000E+000
      The stream velocity in the y direction is  0.000000000000000E+000
  Plane flow
  The minimum x coordinate is  0.000000000000000E+000
  The minimum x coordinate is a solid surface
  The minimum x boundary is a surface with the following properties
      The temperature of the surface is   273.000000000000
      The fraction of specular reflection is  0.000000000000000E+000
      The    velocity   in   the   y   direction   of   this   surface   is
0.000000000000000E+000
  The maximum x coordinate is   200.000000000000
  The maximum x coordinate is a solid surface
  The maximum x boundary is a surface with the following properties
      The temperature of the surface is   273.000000000000
      The fraction of specular reflection at this surface is
      0.000000000000000E+000
      The velocity in the y direction of this surface is   168.595000000000
  The flowfield is initially the stream(s) or reference gas
  There is no secondary stream initially at x > 0
  The desired number of molecules in a sampling cell is         250
```

The steady state velocity distribution between the plates is, as shown in Fig. 7.14 and as expected, very nearly linear with a velocity slip close to $U_p K_n$ at each surface.

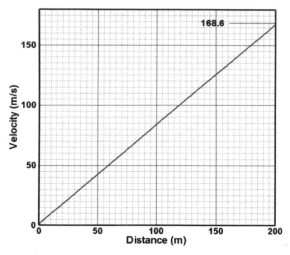

Fig. 7.14 The velocity profile for *s*=0.5 Couette flow in argon.

The speed ratio of 0.5 was chosen in an attempt to minimize compressibility effects while still having a sufficiently large disturbance to not require an excessively large DSMC sample. Fig. 7.15 shows that there is a parabolic temperature profile with a maximum temperature rise of 4.75 K.

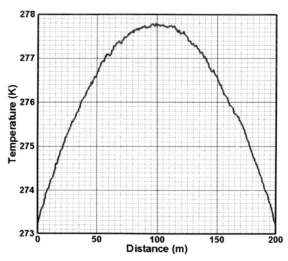

Fig. 7.15 The temperature profile for *s*=0.5 Couette flow in argon.

The computational details are of interest and the first few items in the output file **DS1OUT.DAT** are as follows:

```
DSMC program for a one-dimensional plane flow
 Interval        8434 Time     5978.34529408995        with      1787749
  samples from   1564.40365561119
 Total simulated molecules =       100001
 Species            1  total =        100001
 Total molecule moves   =        1210912509004
 Total collision events =         114006457908
 Species dependent collision numbers in current sample
 0.16835E+12
 Computation time    62094.73      seconds
 Collision events per second   1836008.56420491
 Molecule moves per second   19501050.8863304
 Collision cells =        6400

 Surface at  0.000000000000000E+000
 Incident sample   212225165.000000
 Number flux  1.244671482157390E+020  /sq m/s
 Inc pressure -2.46739805246998E-003  Refl pressure -2.46606473714401E-003
 Pressure -4.933462789613996E-003  N/sq m
 Inc y shear  1.741240870811130E-005  Refl y shear  7.390662327659115E-008
 Net y shear  1.733850208483471E-005  N/sq m
 Net z shear -3.491325521973982E-007  N/sq m
 Incident translational heat flux  0.939782253210386       W/sq m
 Total incident heat flux  0.939782253210386        W/sq m
 Reflected translational heat flux -0.938329241786268       W/sq m
 Total reflected heat flux -0.938329241786268       W/sq m
 Net heat flux  1.453011424117609E-003  W/sq m
 Slip velocity (y direction)  0.813349536395496       m/s
 Translational temperature slip  0.183878224326463       K

 Surface at    200.000000000000
 Incident sample   212226568.000000
 Number flux  1.244679710559943E+020  /sq m/s
 Inc pressure  2.46739330016464E-003  Refl pressure  2.466068087298984E-003
 Pressure  4.933461387463631E-003  N/sq m
 Inc y shear -1.750066470411504E-005  Refl y shear  1.317698458463868E-008
 Net y shear -1.751384168869968E-005  N/sq m
 Net z shear  4.498141480366187E-007  N/sq m
 Incident translational heat flux  0.939733997022958       W/sq m
 Total incident heat flux  0.939733997022958       W/sq m
 Reflected translational heat flux -0.938239506775769       W/sq m
 Total reflected heat flux -0.938239506775769       W/sq m
 Net heat flux  1.494490247188507E-003  W/sq m
 Slip velocity (y direction) -0.818492620391652       m/s
 Translational temperature slip  0.150824715685417       K
```

The calculation was made on a 3 GHz CPU and the computation time turned out to be about ten times higher than the flow time. Almost two million collision events and twenty million molecule move steps were calculated each CPU second. Each collision event involves two molecules so the time step was approximately one fifth the mean collision time per molecule. This is in accordance with the default value of 0.2 that was chosen for the **CPDTM** computational parameter. The work done against the shear stress by the moving upper surface is 0.00295 W/m^2 and this exactly matches the total heat transfer to the surfaces.

While the desired accuracy for the velocity and temperature profiles was readily obtained, the net heat flux is only 0.16% of either the incident or reflected heat flux and extremely large samples are required to accurately determine a quantity that is such a small difference between large quantities. The number molecules that collide with a surface is more than twenty million and, even with very large sample size, the uncertainty in the heat transfer is of the order of ten percent.

The shear stress is not subject to the "small difference between large quantities" difficulty that is associated with the heat transfer. However, because its magnitude in this case is only 0.3% that of the pressure, it is also subject to significant scatter. The velocity gradient is 0.83 s and the average shear stress is 1.75×10^{-5} N/m^2. This corresponds to a viscosity coefficient of 2.11×10^{-5} N s/m^2 and is in good agreement with the nominal value of 2.117 N s/m^2 that was used to determine the effective diameter of the VHS model of an argon molecule.

The magnitudes of the velocity and temperature slips that are sampled, using Eqns. (4.62) and (4.63), from the properties of the molecules incident and reflected from the surface are in good agreement with those extrapolated from the flowfield properties. This facilitates the calculation of velocity and temperature gradients normal to surfaces and, for very low Knudsen numbers where the Chapman-Enskog transport properties are valid and direct samples are excessively noisy, the shear stress and heat transfer can be calculated from the gradients. This is the procedure that can be adopted, especially for the heat transfer, in two and three-dimensional flow calculations when the level of statistical scatter is unacceptable.

This, like most DSMC calculations, has employed the VHS molecular model and the VSS model is recommended only for flows in which diffusion is significant. There is an implicit assumption that there is no difference between the VHS and VSS results when diffusion is not important. The calculation was therefore repeated with the VSS rather than the VHS model. There were no significant differences in the results and the calculation serves to verify the Chapman-Enskog theory on which the molecular parameters are based. The question that then arises is whether the equivalence of VHS and VSS results applies also to large disturbance flows that may be beyond the range of applicability of the Chapman-Enskog theory. The calculation was therefore repeated with a moving wall speed ratio of 4 rather than 0.5.

The VHS and VSS velocity and temperature profiles are compared in Figs. 7.16 and 7.17. The temperature rise is sufficiently large for the higher viscosity coefficient at the centre of the flow to lead to a noticeably lower velocity gradient. The velocity slip increases from 0.81 m/s to 6.1 m/s and scales linearly with the surface speed. On the other hand, the temperature slip increases from about 0.17 K to 9.4 K and, to within the uncertainty associated with the lower value, scales with the square of the surface speed. The most important conclusion is that the results from the VSS model are almost identical to those from the VHS model.

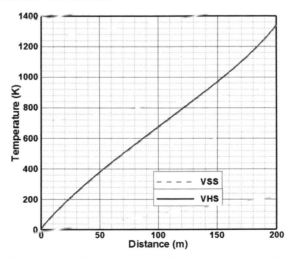

Fig. 7.16 The VHS and VSS velocity profiles for s=4 Couette flow in argon.

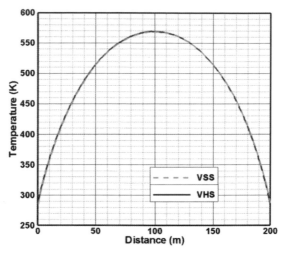

Fig. 7.17 The VHS and VSS temperature profiles for $s=4$ Couette flow.

Couette flow involves only viscous and heat conductivity effects, but the far upstream influence ahead of shock waves appears to be caused by the upstream diffusion of molecules from the shock front. The Mach 8 shock calculation that led to Fig. 7.13 was therefore repeated with VSS instead of VHS molecules. Figure 7.18 shows that the higher self-diffusion coefficient leads to an extension of the upstream influence and confirms that it is diffusive in nature.

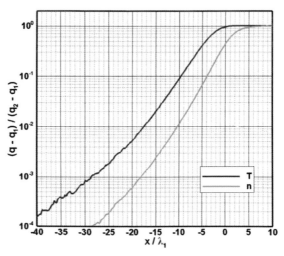

Fig. 7.18 Mach 8 argon shock structure using the VSS molecular model.

7.5 Shock wave structure in a gas mixture

Consider a shock of Mach number 8 in a mixture of helium and xenon with proportions such that the molecular weight is equal to that of argon. The data file for this case is:

```
Data summary for program DSMC
The n in version number n.m is          1
The m in version number n.m is          1
The approximate number of megabytes for the calculation is        10
        7
   Helium-argon-xenon mixture
The gas properties are:-
    The stream number density is  2.422800000000000E+018
    The stream temperature is   273.000000000000
    The stream velocity in the x direction is    2462.52000000000
    The stream velocity in the y direction is   0.000000000000000E+000
    The fraction of species      1  is  0.717760000000000
    The fraction of species      2  is  0.000000000000000E+000
    The fraction of species      3  is  0.282240000000000
Plane flow
The minimum x coordinate is   -50.0000000000000
The minimum x coordinate is a stream boundary
The maximum x coordinate is    50.0000000000000
The maximum x coordinate is a stream boundary with a fixed number of
simulated  molecules
The flowfield is initially the stream(s) or reference gas
There is a secondary stream applied initially at x = 0 (XB(2) must be > 0)
The secondary stream (at x>0 or X>XS) properties are:-
    The stream number density is  9.257210000000000E+018
    The stream temperature is    5698.10000000000
    The stream velocity in the x direction is    644.492100000000
    The stream velocity in the y direction is   0.000000000000000E+000
    The fraction of species      1  is  0.717760000000000
    The fraction of species      2  is  0.000000000000000E+000
    The fraction of species      3  is  0.282240000000000
There is no molecule removal
The desired number of molecules in a sampling cell is          250
```

The downstream boundary is again chosen to conserve the number of simulated molecules and the conservation is applied separately to each molecular species. However, there is a separation of species within the shock wave that alters the overall species number ratio and the conservation cannot be applied until the shock is fully formed. Therefore, in a gas mixture, the maintenance of a constant molecule number to avoid a random walk in the shock position is an option that may be chosen when the program is restarted with a new sample.

The stream number density is again set such that the mean free path in the undisturbed upstream gas is unity. The profiles of the overall gas properties are shown in Fig. 19. While the average molecular weight matches that of argon, the profiles are qualitatively different from the corresponding argon profiles in Fig. 7.10. The extent of the wave is greater by about five men free paths in both the upstream and downstream directions. Because the composition of the gas varies, the number density profile is no longer identical to the density profile and the differences are substantial. In particular, the number density profile is highly asymmetric. On the other hand, the temperature nonequilibrium appears to be much reduced.

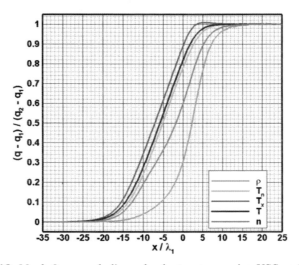

Fig. 7.19 Mach 8 xenon-helium shock structure using VSS molecules.

The overall gas properties that are plotted in Fig. 7.19 have been defined as averages over the molecular species of similar quantities based on the molecules of a single species. The profiles for the separate species are plotted in Fig. 7.20 for helium and in Fig. 7.21 for xenon. The shock Mach number is defined as the ratio of the speed of the wave relative to the undisturbed gas to the speed of sound in this gas. The relative speed in this case would correspond to a shock Mach number of 2.54 in pure helium and 14.51 in pure xenon. It is no surprise that the helium and xenon profiles are characteristic of those in a weak shock and a strong shock, respectively.

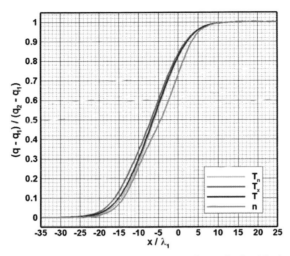

Fig. 7.20 Profiles of the **helium** quantities through the Mach 8 shock.

The temperature rise in helium leads that in xenon by about five upstream mean free paths and the helium density rise is about ten free paths ahead of that in xenon. The species concentrations are constant in the upstream gas and, because the separation effects are a result of diffusion, the VSS molecular model has been employed in the calculation.

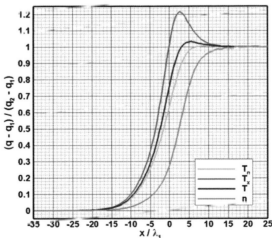

Fig. 7.21 Profiles of the **xenon** quantities through the Mach 8 shock.

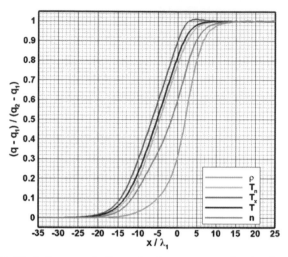

Fig. 7.22 Mach 8 xenon-helium shock structure using VHS molecules.

The difference in molecular weight is extreme in this case and substantial differences were expected between calculations that employ the VHS and VSS molecular models. However, a comparison of the VHS profiles in Fig. 7.22 with the VSS profiles in Fig. 7.19 shows that they are qualitatively similar and the VSS shock thickness is less than 10% larger than the VHS thickness. The differences can be expected to be very small in flows that involve less disparate mixtures and the VHS model may be used almost exclusively

7.6 Structure of a re-entry shock wave

The preceding shock wave calculations have been concerned only with monatomic gases and, rather than progressively add rotational, vibrational and electronic energy and chemical reactions, this example includes all these effects. It is the structure of a stationary normal shock wave in a stream moving at 7,500 m/s in the atmosphere at an altitude of 80 km. Neglecting small fractions of atomic oxygen and argon, the mean CIRA-2012 COSPAR model of the Earth's upper atmosphere predicts 79.7% nitrogen 20.3% oxygen at a number density of 3.77×10^{20} and at a temperature of 198 K. Because of the partial excitation of vibration and electronic energy and the presence of chemical reactions that are slow to reach equilibrium, the conditions downstream of the wave are not known.

The downstream boundary conditions for the calculations of shock wave structure in the preceding sections made use of the Rankine-Hugoniot jump relations. Because it is no longer possible to calculate these relations, an alternative downstream boundary conditions must be applied and it has been set here to a specularly reflecting wall. The data file for this case was listed by **DS1D.TXT** as:

```
Data summary for program DSMC
  The n in version number n.m is          1
  The m in version number n.m is          1
  The approximate number of megabytes for the calculation is          6
        6nn Real air @ 7.5 km/s
  The gas properties are:-
     The stream number density is  3.770000000000000E+020
     The stream temperature is    198.000000000000
     The stream vibrational and electronic temperature is
   198.000000000000
     The stream velocity in the x direction is   7500.00000000000
     The stream velocity in the y direction is  0.000000000000000E+000
     The fraction of species         1  is  0.203000000000000
     The fraction of species         2  is  0.797000000000000
     The fraction of species         3  is  0.000000000000000E+000
     The fraction of species         4  is  0.000000000000000E+000
     The fraction of species         5  is  0.000000000000000E+000
  Plane flow
  The minimum x coordinate is  -1.20000000000000
  The minimum x coordinate is a stream boundary
  The maximum x coordinate is  0.000000000000000E+000
  The maximum x coordinate is a solid surface
  The maximum x boundary is a surface with the following properties
        The temperature of the surface is    1.00000000000000
        The fraction of specular reflection at this surface is    1.00000000
        The velocity in the y direction of this surface is  0.0000000000E+000
  The flowfield is initially the stream(s) or reference gas
  There is no secondary stream initially at x > 0
   Molecule removal is specified whenever the program is restarted
  The desired number of molecules in a sampling cell is          100
```

The "gas model 6" within the **DSMC.F90** source code is called "Real air @ 7.5 km/s" because it has been based on that used in earlier programs that employed the TCE chemistry model. The TCE model employs the chemistry rate equations that were generally accepted to be appropriate at the temperatures behind the bow shock of a blunt body moving in the upper atmosphere at re-entry speed. The program now employs the Q-K theoretical rates that, when using the default settings, apply at all temperatures. However, the Q-K model now permits energy barriers to be specified for both the forward and reverse exchange and chain reactions. It also makes provision for an arbitrary reduction factor to be applied to the dissociation rate.

The auxiliary program **QKrates.exe** permits the Q-K rate coefficients to be compared with the reaction rates that have traditionally been employed for this class of flow. It was found that, after only three adjustments to the default Q-K rates, the two sets of reaction rates are completely compatible. The adjustments were reduction factors of 0.3 to **N_2** dissociation and 0.2 to **NO** dissociation and an energy barrier of $2{\times}10^{-19}$ J to the reaction **$NO+O{\to}O_2+N$**.

After the program is started, a shock wave forms at the downstream boundary and moves back into the gas. However, there is no point in trying to sample this shock wave in a frame of reference moving with this wave. Because it is moving into the stream, it is a stronger shock than the desired one and, without knowledge of the flow conditions downstream of the shock, it is not possible to calculate the stream speed that would lead to a shock of the desired strength. Moreover, it may require a prohibitively long time for the reactions to reach equilibrium and for the shock to reach a constant speed.

The solution to these problems is, at some time after the start of the calculation, to remove molecules near the downstream wall such that the removed mass flow exactly balances the mass inflow across the upstream boundary. It has been shown (Bird, 1994, §12.12) that, if the molecules in a non-reacting flow are removed with probability proportional to their velocity component normal to the flow direction, momentum and energy are also exactly conserved. The data file chooses the option that allows the amount and location of molecule removal to be specified whenever the calculation is restarted.

The calculation started with 60,000 simulated molecules and was stopped when the number reached one million and the shock wave was at -0.83 m. It was restarted with molecule removal equal to the stream inflow number and with the removal uniformly distributed between $x = -0.05$ m and the wall at the origin. The molecule removal initiated an expansion wave, but the shock wave moved further towards the upstream boundary before it was weakened by the expansion wave. The shock wave reached -1.08 m before moving further back into the computational region and, after well damped oscillations, it eventually became steady with the centre of the density profile at -0.86 m, as shown in Fig. 7.23. The overall degree of dissociation decreased and the total number of simulated molecules in the steady state fell to approximately 940,000. The plotted results are based on a sample over the flow time interval from 8 to 28 ms.

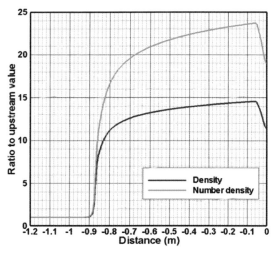

Fig. 7.23 The density and number density profiles.

Both ratios fall sharply in the molecule removal zone. The flow downstream of the shock is subsonic and there is a possibility that the whole downstream flow is affected by this boundary condition. The calculation was repeated with molecule removal being commenced at 667,000 total simulated molecules. A comparison of Figs. 7.24 and 7.23 provides reassurance that the "molecule removal" boundary condition does not affect the flow upstream of the removal zone.

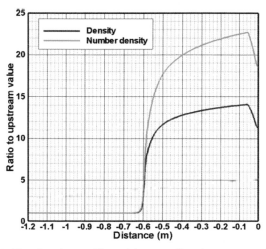

Fig. 7.24 The density profiles with a smaller downstream flowfield.

There is still a significant density gradient at 0.8 m behind the shock wave. This indicates that the dissociation-recombination reactions have not reached equilibrium and this is confirmed by the species fraction profiles in Fig. 7.25. Chemical equilibrium occurs when the recombination rate eventually becomes equal to the dissociation rate. Three-body recombinations are rare events at the densities associated with this flow. The computation involved ten million simulated dissociations, but only seven thousand simulated recombinations. Equilibrium cannot be expected to be attained at less than a distance of the order of 10 m downstream of the wave.

The upstream mean free path is 0.0035 m and the shock wave thickness based on the maximum density gradient over the initial density rise to ten times the freestream density is ten mean free paths. On the other hand, the distance upstream to a one part in a million concentration of atomic species is almost a hundred mean free paths. Nitric oxide also diffuses a significant distance upstream of the maximum density gradient and additional nitric oxide is produced well upstream of the shock by the exothermic reaction $O_2+N \rightarrow NO+O$. Precursor thermal radiation can be expected to extend to at least this distance upstream of the maximum density gradient.

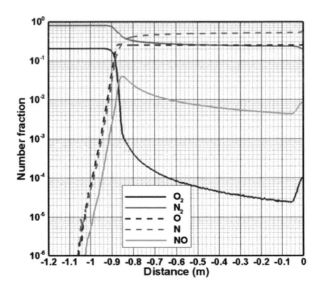

Fig. 7.25 The species number fraction profiles.

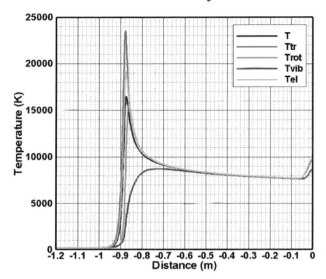

Fig. 7.26 The modal temperature profiles.

The profiles of the modal temperatures across the flowfield in Fig. 7.26 verify that all temperatures come to equilibrium downstream of the shock.   The width of the region of marked temperature non-equilibrium is not much over 100 upstream mean free paths.   With the exception of the inclusion of electronic energy, these results are consistent with those from earlier studies of similar flows.

It should be emphasised that the gas model includes only electrically neutral particles and the electronic temperature is not the temperature of electrons.   Instead, it reflects the fact that many molecules are in electronic states above the ground state that have energy levels below the ionization energy.   This energy has habitually been ignored in previous studies, both continuum CFD and DSMC, of hypersonic flows in air.

The practical problem associated with the inclusion of electronic energy is that there is very little data for the excitation cross-sections. Because of this, an arbitrary "electronic excitation collision number" has been set to fifty for all levels and for all gas species.   This is ten times the rotational relaxation collision number and it is surprising that the electronic temperature rise occurs upstream of the rotational temperature rise.   This is an artifact of a fundamental difference between the temperature definitions for fully excited energy modes and the temperature definitions for partially excited modes.

Because the translational and rotational energies are regarded as fully excited, the temperatures of these modes can be unambiguously based on the average energies in these modes. On the other hand, the temperatures of the vibrational and electronic modes have been based on the ratios of the average energies in these modes to the energies that correspond to equilibrium at the translational temperature. These definitions ensure that the modes have the same temperature at equilibrium and indicate whether or not the energies are above or below the equilibrium values. However, the magnitudes of the reported temperatures may be misleading. If, for example, there are just a few excited molecules in a gas at a low translational temperature, there will be a high modal temperature that is based on very little energy. The modal temperatures do not, therefore, provide a reliable guide to the partitioning of energy between the modes and Fig. 7.27 presents a direct plot of the modal energies.

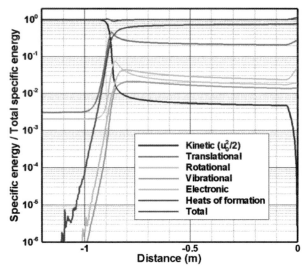

Fig. 7.27 The profiles of modal specific energies.

The rotational and electronic profiles are now in the expected order, but the most striking result is that the electronic energy exceeds both the rotational and vibrational energies in the downstream gas. This reflects the declining fractions of diatomic molecules and the high electronic energy associated with atomic nitrogen. At equilibrium, the energy required to dissociate the diatomic molecules will amount more than 70% of the total energy.

The examples in the preceding sections have been undemanding in terms of computer resources and the criteria for a good DSMC calculation have been met by wide margins. The main criterion is that the ratio of the mean separation of the collision pairs to the mean free path, or mcs/mfp ratio, should be small compared with unity. With the default fifteen molecules per collision cell and without cell adaption, the mcs/mfp ratio in this case was slightly above 0.15 over much of the flow. This is a marginal value and the number of molecules per collision cell was reduced to ten in order to bring it close to 0.1.

The sample size in the flow behind the shock wave was more than 5×10^8. The standard deviation of the statistical scatter was less than one part in ten thousand and there is noticeable scatter in Fig. 7.27 when the energy ratio is less than this value. The scatter becomes severe at the "one part in a million" level.

7.7 Shock formation in air

Sections 7.3, 7.5 and 7.6 have been concerned with the structure of steady normal shock waves. The first two studies were made with unphysical boundary conditions that made use of the Rankine-Hugoniot jump conditions. The re-entry shock wave in §7.6 was initially generated by the normal impact of a uniform stream on a specular surface, but the unsteady phase of the flow was not studied.

The flow to be considered in this section is the generation, by the impact of a stream on a specular end-wall, of a shock of Mach number 2 in air at a number density of 1×10^{20} and a temperature of 300 K. The data file for this case was listed by **DS1D.TXT** as:

```
Data summary for program DSMC
  The n in version number n.m is            1
  The m in version number n.m is            1
  The approximate number of megabytes for the calculation is        1000
          5
  Ideal air
  The gas properties are:-
    The stream number density is  1.000000000000000E+022
    The stream temperature is    300.000000000000
    The stream velocity in the x direction is   436.250000000000
    The stream velocity in the y direction is  0.000000000000000E+000
    The fraction of species        1  is  0.200000000000000
    The fraction of species        2  is  0.800000000000000
  Plane flow
```

```
The minimum x coordinate is -2.000000000000000E-003
The minimum x coordinate is a stream boundary
The maximum x coordinate is  0.000000000000000E+000
The maximum x coordinate is a plane of symmetry
The flowfield is initially the stream(s) or reference gas
There is no secondary stream initially at x > 0
There is no molecule removal
 The outer boundary is stationary
The desired number of molecules in a sampling cell is       100000
```

The flow extends from -0.002 m to the origin and, at zero time, it is filled by the gas of simulated air molecules with a stream speed of 436.25 m/s. This is the speed that, when impulsively stopped by a specular wall, results in a Mach 2 shock wave in a diatomic gas in which the speed of sound is 349 m/s. The mean free path in the undisturbed gas is 0.0001275 m. The generation of a shock wave by a moving piston in a monatomic gas is a similar flow and has been studied by Bird (1994). The mass-centre of the wave was almost coincident with the path of the discontinuous continuum shock and the downstream Rankine-Hugoniot conditions were attained as soon as the shock was fully formed. There was then a uniform gas between the shock wave and the piston. The addition of the rotational mode can be expected to lead to a more complicated flow and particular attention will be paid to the ratio of the translational temperature to the rotational temperature.

Before studying the unsteady flow, it is desirable to obtain the profiles of the macroscopic properties in the steady shock wave through a data file similar to that employed in §7.3 and §7.5. The temperature and number density profiles are shown in Figs. 7.28 and 7.29, respectively. The temperature non-equilibrium is confined to a distance interval of 0.0028, or 28 upstream mean free paths. The rotational relaxation collision number is set to five for both oxygen and nitrogen in the "ideal air" gas model that has been employed. There is provision in the gas data specification for this number to be set as a first order polynomial in temperature and there is some data for the temperature dependence. The data of Parker (1959) suggests a value of about four at 300 K and five at 500 K. On the other hand, the data of Lordi and Mates (1976) suggests five at 300 K and seven at 500 K. Given that the correct value in uncertain and the evidence from Fig. 7.28 is that the effect of this value in confined to the internal structure of the wave, there seems little point in using other than the nominal value of five.

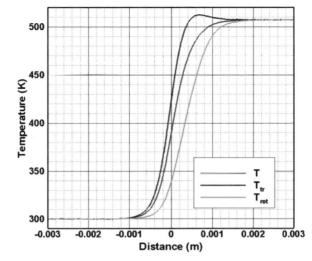

Fig. 7.28 Temperature profiles through a Mach 2 shock wave in air.

It has generally been found that the rotational temperature profile in a shock wave is very close to the density profiles. A comparison of the profiles in Figs. 7.28 and 7.29 shows that, in this case, the rotational temperature profile is slightly ahead of the density profile.

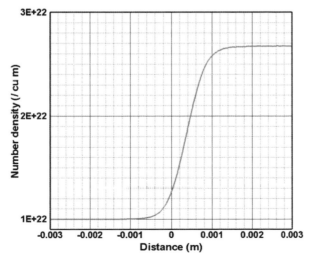

Fig. 7.29 Number density profile through a Mach 2 shock wave in air.

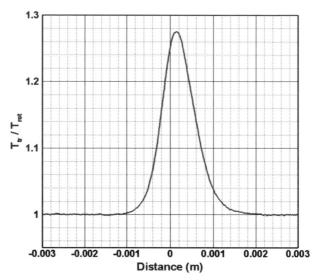

Fig. 7.30 Translational-rotational non-equilibrium within the wave.

The ratio of the translational to the rotational temperature in the steady shock wave that is shown in Fig. 7.30 may be compared with the corresponding ratio during the formation of the wave.

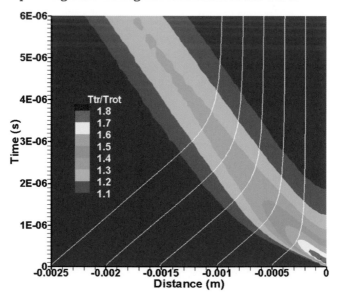

Fig. 7.31 Non-equilibrium during the formation of the wave.

The intercept between the increasing and decreasing profiles at a T_{tr}/T_{rot} ratio of 0.1 in the steady shock of Fig. 7.30 is very close to 1mm.   This matches the intercept between the corresponding contours in Fig 7.31 after the shock is formed and ratio has a maximum of 1.27. Immediately after the specular wall at the origin is inserted at zero time, the T_{tr}/T_{rot} ratio has the much higher value of 1.87. Apart from the delayed temperature increase at the origin, the overall temperature contours in Fig. 7.32 are similar to those (Bird, 1994) for a monatomic gas. While the earlier studies were made as the ensemble average of six hundred separate runs, these contours are based on a single run. The initial number of simulated molecules was ten million and this increased to more than 20 million during the run. The sample size with a resolution of 100 sampling cells was then five to ten million. The automatically set initial time step is generally too large and it is reset by the program at the end of the first output interval. This is unsuitable for the generation of x-t diagrams and there is a line in the source code that, coupled with the fixed time step option in the computational parameters, allowed the time step to be set such that the resolution is similar in both the x and t directions.

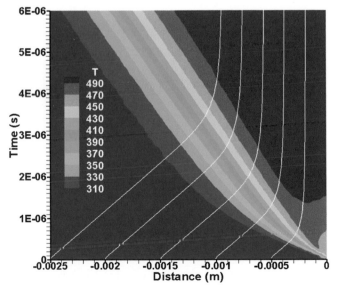

Fig. 7.32 The overall temperature (K) in the distance-time plane.

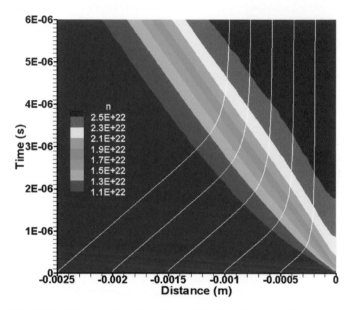

Fig. 7.33 The number density (m⁻³) in the distance-time plane.

The number density profiles in Fig. 7.33 are also similar to those in Bird (1994). There is initially a near doubling of the number density and this corresponds to the collisionless result. The midpoint of the density rise is at 1.83×10^{22} which is near the centre of the brightest green contour interval. The discontinuous continuum shock path is within this contour interval.

The Navier-Stokes equations are generally thought to be valid for the prediction of shock structure up to a shock Mach number of about 1.8 and should not be in error by more than about five percent for a Mach two wave. Some degree of pressure tensor asymmetry in inherent in the viscosity coefficient, but the equations would miss qualitative features such as the non-equilibrium between the translational and rotational modes. Also, unless they were extended to include a term for the pressure diffusion, they would fail to predict the species separation that occurs in a gas mixture. In this case, the nitrogen number fraction increases by 0.32% at the location of the maximum translational-rotational non-equilibrium. This corresponds to a 1.28% depletion of oxygen at this location.

7.8 Spherically imploding shock wave

The examples in Bird (1994) included a spherically imploding shock wave generated by a spherical specularly reflecting piston at a radius of 1 m that, at zero time, acquired a velocity towards the origin of -2285.5 m/s. The initially uniform argon inside the piston was at a temperature of 273 K and a number density of 1×10^{20} m⁻³ and a plane piston at this speed would produce a shock of Mach 10. However, the initial number of mean free paths between the origin and piston is only about 70 and a significant fraction of the flowfield was occupied by the internal structure of the shock wave. The case is repeated in this section for an initial number density of 5×10^{20} m⁻³ so that the shock waves are more clearly defined. The data file for this case was listed by **DS1D.TXT** as:

```
Data summary for program DSMC
 The n in version number n.m is          1
 The m in version number n.m is          1
 The approximate number of megabytes for the calculation is          400
          2       Argon
 The gas properties are:-
     The stream number density is  5.000000000000000E+020
     The stream temperature is   273.000000000000
     The stream velocity in the x direction is  0.000000000000000E+000
     The stream velocity in the y direction is  0.000000000000000E+000
 Spherical flow
 The minimum x coordinate is  0.000000000000000E+000
 The minimum x coordinate is an axis or center
 The maximum x coordinate is   1.00000000000000
 The maximum x coordinate is a plane of symmetry
 There are no radial weighting factors
 The flowfield is initially the stream(s) or reference gas
 There is no secondary stream initially at x > 0
 The outer boundary moves with a constant speed
 The speed of the outer boundary is  -2285.50000000000
 The desired number of molecules in a sampling cell is          50000
```

The temperature and number density contours in the r-t plane up to the time that the shock wave just reaches the origin are shown in Figs. 7.34 and 7.3, respectively. The temperature behind a plane normal shock wave in the undisturbed gas is 8775 K and the number density is 1.94×10^{21} m⁻³. This temperature is attained almost immediately after the piston starts to move and the number density at the piston reaches this value before 0.00003 s. The temperature behind the shock at the start of the shock-axis interaction exceeds 55,000 K and this corresponds to a shock Mach number just over 54.

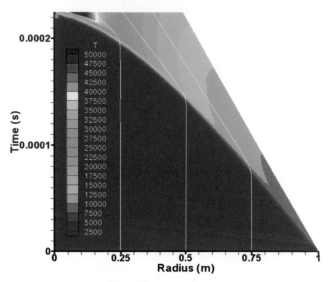

Fig. 7.34 The temperature (K) up to the shock-centre interaction.

There is a fundamental qualitative difference between the temperature and number density contours in that the maximum temperature is immediately behind the wave and the maximum density is at the piston.

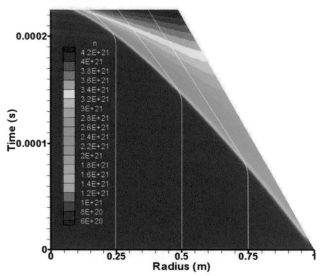

Fig. 7.35 The number density (m⁻³) up to the shock-centre interaction.

The calculation assumes that argon is a perfect gas and there would be significant ionization in real argon. This would affect the maximum temperature that would be far lower than the 50,000 K in Fig. 7.34. This remark applies with even greater force to the temperatures that are involved in the reflection of the shock from the origin of the spherical flow. As shown in Fig. 7.36, the ideal gas temperature at the centre approaches 300,000 K. The corresponding temperature in the earlier calculation at one fifth the number density was approximately 150,000 K.

The calculation was made without weighting factors and the "desired number of molecules in a sampling cell" relates to a mean value. The sample size in the outermost of the 400 uniform width sampling cells at a time of 0.00023 s was five million while that in the innermost cell was only 330. The next cell contained only a thousand so that there is a large degree of scatter at the centre and this is evident in Fig. 7.36. While this is undesirable, random walks associated with weighting factors would lead to larger and less predictable spurious effects at the centre of the flow. The measures to enforce a uniform initial flow in **DSMC.F90** have avoided the difficulties that were encountered in the earlier calculation (Bird, 1994).

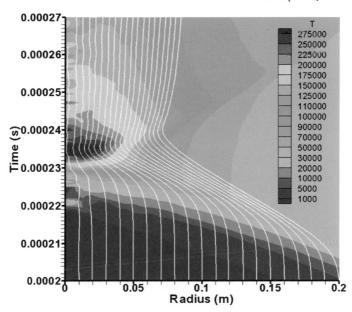

Fig. 7.36 The temperature (K) as the shock reflects from the origin..

Following the first centre interaction that is shown in Fig. 7.36, the shock wave continues to reflect between the contracting spherical piston and the centre. Figure 7.37 shows the compression process up to the fifth interaction of the shock wave with the centre of the sphere. The ideal gas temperature at the centre just reaches one million K at the second interaction and most of the flow is near this temperature by the time of the fifth interaction. However the temperature does not increase much beyond one million degrees. This is because the piston speed is then small in comparison with the speed of sound in the gas and the shock wave becomes weak.

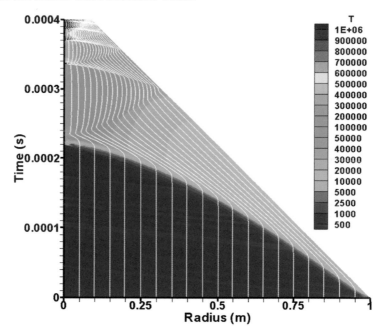

Fig. 7.37 The temperature (K) in multiple reflections of the shock wave.

The centre of the sphere is a singularity in a continuum model that treats a shock wave as a discontinuity and early studies of the problem were limited to similarity methods that largely ignored the physics. The singularity disappears when the shock wave has a finite thickness. Real gas effects should, of course, be included in serious studies of imploding shock waves. In addition, a spherical piston with a uniform velocity towards the centre is not necessarily the most appropriate outer boundary condition.

7.9 Formation of a combustion wave

The spontaneous and forced combustion of a stoichiometric mixture of oxygen and hydrogen was considered in Sections 6.6 and 6.7 as a homogeneous gas problem. A related one-dimensional problem is the ignition at one end of a finite extent of gas. This forms a combustion wave that moves into the unburned gas. The data file for this case was listed by **DS1D.TXT** as:

```
Data summary for program DSMC
  The n in version number n.m is            1
  The m in version number n.m is            1
  The approximate number of megabytes for the calculation is         400
          8      Oxygen-hydrogen
  The gas properties are:-
     The stream number density is  1.343500000000000E+024
     The stream temperature is   15000.0000000000
     The stream vibrational and electronic temperature is
    15000.0000000000
     The stream velocity in the x direction is  0.000000000000000E+000
     The stream velocity in the y direction is  0.000000000000000E+000
     The fraction of species         1  is  0.666700000000000
     The fraction of species         2  is  0.000000000000000E+000
     The fraction of species         3  is  0.333300000000000
     The fraction of species         4  is  0.000000000000000E+000
     The fraction of species         5  is  0.000000000000000E+000
     The fraction of species         6  is  0.000000000000000E+000
     The fraction of species         7  is  0.000000000000000E+000
     The fraction of species         8  is  0.000000000000000E+000
  Plane flow
  The minimum x coordinate is -1.000000000000000E-005
  The minimum x coordinate is a plane of symmetry
  The maximum x coordinate is  1.000000000000000E-004
  The maximum x coordinate is a stream boundary
  The flowfield is initially the stream(s) or reference gas
  There is a secondary stream applied initially at x = 0 (XB(2) must be > 0)
  The secondary stream (at x>0 or X>XS) properties are:-
     The stream number density is  2.687000000000000E+025
     The stream temperature is   750.000000000000
     The stream vibrational and electronic temperature is
    750.000000000000
     The stream velocity in the x direction is  0.000000000000000E+000
     The stream velocity in the y direction is  0.000000000000000E+000
     The fraction of species         1  is  0.666700000000000
     The fraction of species         2  is  0.000000000000000E+000
     The fraction of species         3  is  0.333300000000000
     The fraction of species         4  is  0.000000000000000E+000
     The fraction of species         5  is  0.000000000000000E+000
     The fraction of species         6  is  0.000000000000000E+000
     The fraction of species         7  is  0.000000000000000E+000
     The fraction of species         8  is  0.000000000000000E+000
  The desired number of molecules in a sampling cell is         10000
```

The unburned gas was at a temperature of 750 K and standard number density of 2.687×10^{25} m^{-3}. The reaction was initiated through a small region extremely hot gas between -1×10^{-5} m and the origin. The temperature was increased by a factor of twenty to 15,000 K and the number density was reduced by a similar factor so that the initial gas was at a uniform pressure with 0.5% of the molecules at the high temperature. Fig. 7.38 shows that the water vapour fraction gradually increased at the upstream plane of symmetry and eventually reached a value of just over 0.6. When the extent of the hot gas was halved, the water vapour fraction increased to a value of about 0.3 and subsequently declined. There is evidently a minimum amount of added energy for sustained ignition.

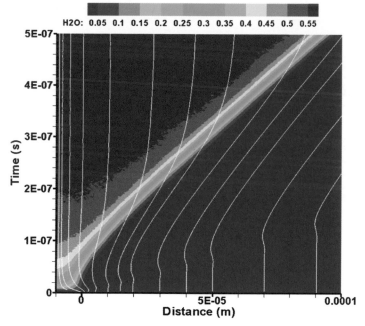

Fig. 7.38 The water vapour fraction during the formation of the wave.

The combustion wave speed quickly settled down to a constant value, but there was a thinning of the wave over a much longer time. The final number of upstream mean free paths between the first and last contours is about forty, but the full width of the wave is well over a hundred upstream mean free paths.

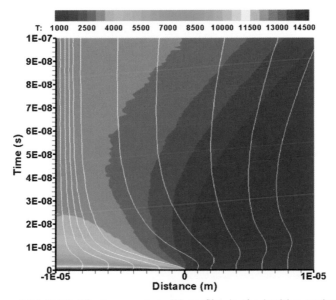

Fig. 7.39 The temperature (K) profiles in the ignition region.

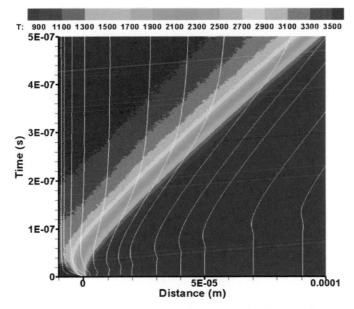

Fig. 7.40 The temperature (K) contours in the overall flow.

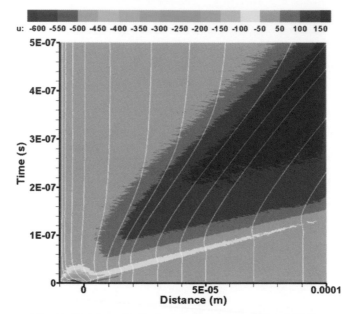

Fig. 7.41 The velocity (m/s) contours in the overall flow.

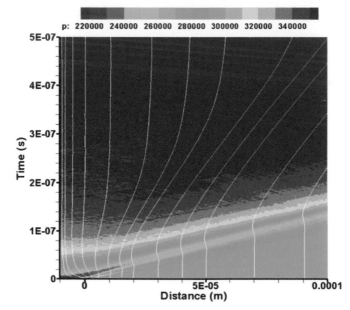

Fig. 7.42 The pressure (N/m²) profiles in the overall flow.

Figure 7.39 shows the temperature contours in the ignition region, while Fig. 7.40 shows the contours of the much lower temperatures in the overall flow.   Note that the spacing of the first and last temperature contours in the combustion wave is much greater than the spacing of the water vapour contours in Fig. 7.38.   The temperature contours provide a more accurate indication of the width of the combustion wave.   The initial 15,000 K drops rapidly and halves within five nanoseconds. This causes the density to increase and the initially very hot gas contracts towards the plane of symmetry. The velocity contours in Fig, 7.41 show that the magnitude of the induced negative velocity exceeds 600 m/s. The resulting expansion wave extends across the flowfield and is visible in both the velocity contours and the pressure contours.   The pressure contours show that the expansion wave is preceded by a weaker compression wave that results from the initial interaction of the very hot molecules with the undisturbed gas molecules. Both these waves are clearly visible in the particle paths. The magnitude of the velocity changes decreases from path to path, so that the acoustic waves decay appreciably by the time that they reach the downstream flow boundary. These initial waves originate at the origin and travel with the speed of sound in the unburned gas. This is approximately 850 m/s, and that the flowfield is traversed in about 120 nanoseconds. The downstream boundary is a stream boundary and, because the entering molecules do not match the molecules that leave across this boundary, boundary interference effects can be expected. These appear to be surprisingly small.

The weak waves from the ignition process are followed by the compression waves that are produced by the accelerating flame front. The gradual convergence of these waves is visible in Fig, 7.42 and they would eventually coalesce at about 1 mm to form the pre-combustion shock wave. The flow velocity behind these waves and in front of the fully formed combustion wave is approximately 160 m/s. The speed of the combustion wave is 220 m/s so that its speed relative to the unburned gas is 60 m/s.   For combustion at room temperature, the combustion wave speed relative to the unburned gas would be only a few meters per second and the width of the combustion wave would be of the order of 100,000 upstream mean free paths.

Figure 7,42 shows that the combustions essentially takes place at constant pressure while the combustion in a homogeneous gas that was studied in §6.6 and §6.7 took place at constant volume.

7.10 Non-equilibrium in a spherical expansion

The breakdown of continuum flow in gaseous expansions occurs when the collision rate falls below the rate that is required for the maintenance of the temperature drop that is predicted by isentropic theory. It has been shown (Bird, 1994) that the breakdown is marked by the parallel temperature ceasing to fall at the predicted rate while the normal temperature continues to fall at near the predicted rate. The parallel temperature is constant in the collisionless limit, but the normal (kinetic) temperature falls indefinitely as the molecular velocities become more nearly parallel. The velocity of the gas reaches a limiting value and the density is then inversely proportional to the area of the flow.

The breakdown of the continuum flow assumption is generally associated with low densities, but can occur at high densities if the flow dimensions are sufficiently small. This can be illustrated by the flow in very small supersonic nozzles in which the expansion is very nearly spherical in nature. The first case is the expansion of nitrogen where the conditions just downstream of the throat at a radius of 0.0001 m are standard number density, a temperature of 300 K and a velocity of 400 m/s. The calculation is terminated at a radius of 0.001 m, so that the maximum to minimum area ratio is 100. The data file for this case was listed by **DS1D.TXT** as:

```
Data summary for program DSMC
  The n in version number n.m is          1
  The m in version number n.m is          1
  The approximate number of megabytes for the calculation is        100
          3    Ideal nitrogen
  The gas properties are:-
      The stream number density is  2.686780000000000E+025
      The stream temperature is    300.000000000000
      The stream velocity in the x direction is   400.000000000000
      The stream velocity in the y direction is  0.000000000000000E+000
  Spherical flow
  The minimum x coordinate is  1.000000000000000E-004
  The minimum x coordinate is a stream boundary
  The maximum x coordinate is  1.000000000000000E-003
  The maximum x coordinate is a vacuum
  There are no radial weighting factors
  The flowfield is initially a vacuum
  There is no secondary stream initially at x > 0
  The desired number of molecules in a sampling cell is        2500
```

The initial state of the flow is a vacuum and the average density in eventual steady flow is small in comparison with the stream density. The automatic calculation of the number of real molecules that are represented by each simulated molecule is based on a uniform flow at the stream density. Therefore, in order to obtain the expected number of molecules, the adjustable computational parameter **FNUMF** in the source code was assigned the value 0.01 in place of the default 1. The initial mean free path is approximately 4.7×10^{-8} m.

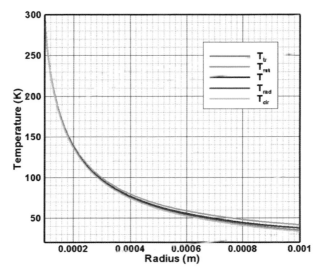

Fig. 7.43 Temperatures in a spherical expansion from 0.0001 to 0.001 m.

All the temperatures are within one degree K when the radius is less than 0.2 mm and, even at the maximum radius, the rotational temperature is only 6.7 degrees higher than the translational temperature of 34.85 K. The radial and circumferential temperatures at the end of the expansion were 37.65 K and 33.45 K, respectively. Figure 7.44 shows that the degree of non-equilibrium is greatly increased if the linear dimensions are reduced by a factor of ten. In both cases, the separation of the rotational and translational temperatures is less than twice the separation of the radial and circumferential temperatures. The region of translational-rotational non-equilibrium is almost identical to the region of translational non-equilibrium.

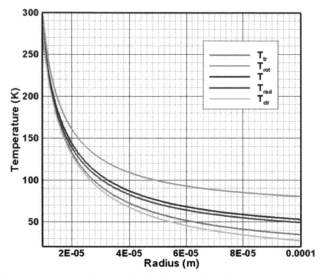

Fig. 7.44 Temperatures in a spherical expansion from 0.00001 to 0.0001 m.

The greater the degree of non-equilibrium, the higher the overall temperature in the flow, so that the Mach number gradient in the flow is reduced. This effect is illustrated in Fig. 7.45.

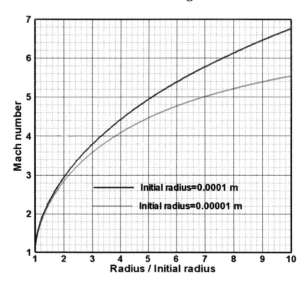

Fig. 7.45 The Mach number profiles in the two cases.

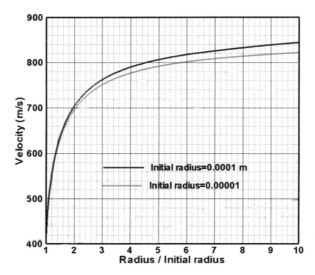

Fig. 7.46 The velocity profiles in the two cases.

Figure 7.46 shows that the flow velocity is less affected than the Mach number. The reduction in flow velocity is at the exit is 2.5% compared with more than 18% for the Mach number. Because the area ratios are the same in each case, the density at the exit is inversely proportional to the velocity.

The vibrational relaxation collision number at low temperatures is many orders of magnitude larger than the rotational relaxation collision number and it is difficult to simulate an expansion in which there is a significant reduction in the vibrational temperature. The test case involves oxygen initially at standard number density and at a temperature of 10,000 K, but with **KREAC=0** so that dissociation and recombination was suppressed. Even with the very high temperature, it was necessary to increase the size of the flowfield by setting the initial and final radii to 0.001 and 0.01 m, respectively. The Knudsen number based on the difference in radius and the mean free path on the entry gas is 0.000015. The initial velocity was 2,000 m/s and this corresponds to an entry Mach number of 1.05.

Fig. 7.47 shows that the vibrational temperature drifts above the other temperatures right from the initial radius and freezes at a temperature of 7,270 K. The translational non-equilibrium barely shows in the plot, but the radial temperature at the exit is 28 K above the translational temperature and the circumferential temperature is

14 K below. The rotational temperature at the exit is 81 K above the translational temperature so that the ratio of rotational to translational non-equilibrium is larger than in the earlier examples.

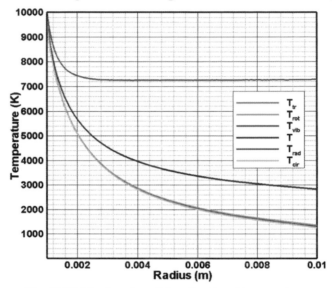

Fig. 7.47 The freezing of the vibrational temperature.

The maximum Mach number of 5.2 in this case is less than that in the previous two cases. However, the reported value is based on the overall temperature that is influenced by the high vibrational temperature and the effective number of vibrational degrees of freedom. The speed of sound is not affected by frozen modes and it would be more realistic to base the Mach number on the near equal translational and rotational temperatures and the assumption of a gas with two degrees of freedom.

Because of the assumption of the simple harmonic model of vibration, it has been possible define a vibrational temperature that leads to realistic values of this quantity when the translational and rotational temperatures are very much lower. There would have been problems with the electronic temperature because it has been defined as the translational temperature multiplied by the ratio of the electronic energy to the equilibrium electronic energy and it would be unrealistically large if the electronic energy became frozen in an expansion. There would be similar difficulties with the definition of the vibrational temperature if anharmonic levels were employed.

7.11 Spherical blast wave

Consider a spherical shock tube with both the high and low pressure gases at a temperature of 300 K and with an initial pressure ratio of ten. The high pressure section extends from the origin to a radius of one meter and the number density of the hard sphere gas is such that the mean free path in this gas is one millimeter. The data file for this case was listed by **DS1D.TXT** as:

```
Data summary for program DSMC
 The n in version number n.m is            1
 The m in version number n.m is            1
 The approximate number of megabytes for the calculation is        2000
           1      Hard sphere gas
 The gas properties are:-
     The stream number density is   1.406740000000000E+021
     The stream temperature is    300.000000000000
     The stream velocity in the x direction is   0.000000000000000E+000
     The stream velocity in the y direction is   0.000000000000000E+000
 Spherical flow
 The minimum x coordinate is   0.000000000000000E+000
 The minimum x coordinate is an axis or center
 The maximum x coordinate is    5.00000000000000
 The maximum x coordinate is a stream boundary
 There are no radial weighting factors
 The flowfield is initially the stream(s) or reference gas
 There is a secondary stream
 The secondary stream boundary is at r=   1.00000000000000
 The secondary stream (at x>0 or X>XS) properties are:-
     The stream number density is   1.406740000000000E+020
     The stream temperature is    300.000000000000
     The stream velocity in the x direction is   0.000000000000000E+000
     The stream velocity in the y direction is   0.000000000000000E+000
 The desired number of molecules in a sampling cell is        50000
```

Note that the calculation was made without weighting factors and with no cell adaption so that, even though it employed 20 million simulated molecules, the sample in the innermost cells is very small. The startup option for continuing unsteady flow sampling must be chosen and there is noticeable statistical scatter in the flow property contours in the distance-time plots of Figs. 7.48 to 7.53.

The shock wave, the contact surface and the expansion waves up to their interaction with the end of the high pressure section would all have a constant speed in a plane flow. In the spherical flow, it is only the leading edge of the expansion wave that has a constant speed. Like the shock wave, the contact "surface" has a finite thickness in a real flow and the temperature contours of Fig. 7.48 show that, in the very early stages of the flow, the width of the contact surface is similar to that of the shock wave.

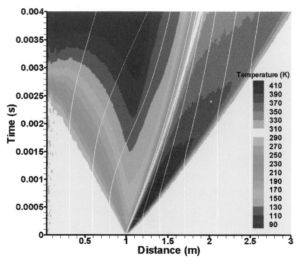

Fig. 7.48 The temperature contours in the early stages of the flow.

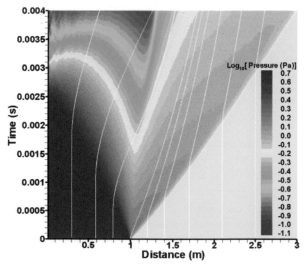

Fig. 7.49 The pressure contours in the early stages if the flow.

The most notable feature of the flow at larger times is the implosion that follows the over-expansion in the early stages of the flow.   Fig. 7.50 shows that the particle path from the initial discontinuity at unit radius expands only to a radius of 2 m before contracting.   The concave particle paths give rise to compression waves that coalesce to form a converging shock wave that starts at a radius of 1.3 m and strengthens as its radius decreases. The extent of the over-expansion is such that the temperature at the origin drops to about 30 K before increasing to more than 700 K behind the reflected shock. The temperature rise associated with the imploding shock in the over-expanded high pressure gas is more than five times that associated with the blast wave in the low pressure gas. Moreover, it was shown in §7.8 that the maximum temperature associated with a spherically imploding shock wave increases as the Knudsen number of the flow decreases and, for denser flows, the maximum temperature could be many times higher. The flow velocity behind the reflected shock is small and there will not be more than one cycle of strongly expanding and contracting flow. The outer boundary is a stream boundary and the exiting molecules with a small outward stream velocity combine with the entering molecules that have zero stream velocity. This leads to a very weak reflected compression wave that shows only as a kink in the temperature contours.

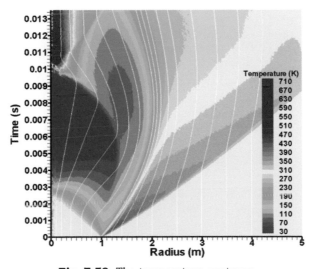

Fig. 7.50 The temperature contours.

The corresponding pressure, number density and velocity contours are shown in Figs. 7.51, 7.52 and 7.53, respectively. An isothermal gas with an initial density and pressure discontinuity is a simple boundary condition and the qualitative differences between the plane and spherical flow cases are drastic. While the magnitude of the temperature rise in the spherical flow case is far higher than that in the plane flow cases, the pressures and densities are everywhere less than their values in the initial high pressure section. The Knudsen number defined by the length of the high pressure section and the mean free in that section case is sufficiently low for the overall wave structure to be well defined. The thickness of the shock waves and the contact surface will change in inverse proportion to the gas density. Also, as noted earlier, the strength of the imploding shock wave when it reaches the origin is a function of this Knudsen number.

In all of these figures, the reflected shock wave is more clearly defined than in the Fig. 7.50 temperature contours. The pressure ratio across the combined incident/reflected implosion wave is approximately five hundred. The reflection of the implosion wave at the origin occurs after 10 milliseconds and there are then two weak shock waves of similar initial strengths moving away from the origin. An observer at a large distance from the origin would hear two strong acoustic waves separated by rather more than ten milliseconds.

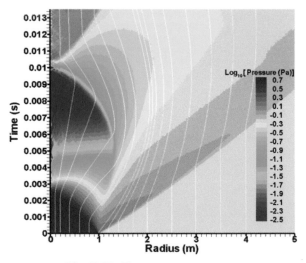

Fig. 7.51 The pressure contours.

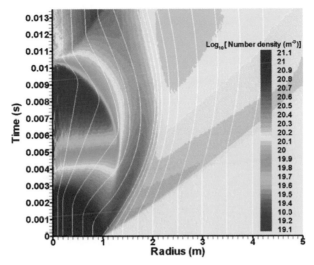

Fig. 7.52 The number density contours.

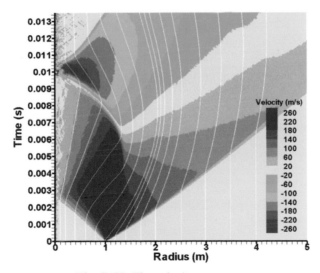

Fig. 7.53 The velocity contours.

The scatter in the velocity contours is particularly marked near the origin because the velocities are small in that region. There are small bumps in the velocity contours at the contact surface and also at the weak reflected compression wave from the outer flow boundary.

7.12 Gas centrifuge

The preceding examples in this chapter have been for either plane or spherical flows. The source code contains statements that are operative only in cylindrical flow and it is necessary to have a cylindrical flow test case. A gas centrifuge was one of the least satisfactory test cases in Bird (1994) and, because the computer that is used for these examples is five hundred times faster than that employed for the earlier examples, a gas centrifuge has again been chosen as the test case.

The first test case is the flow inside a cylinder of radius 0.1 m that starts rotating at zero time with a peripheral speed of 500 m/s. The gas is a mixture of equal parts by number of helium, argon and xenon. The initial number density in the initially uniform gas is such that the radius is approximately 100 mean free paths and the temperature is 300 K. This is also the temperature of the diffusely reflecting surface. The data file for this case was listed by **DS1D.TXT** as:

```
Data summary for program DSMC
  The n in version number n.m is            1
  The m in version number n.m is            1
  The approximate number of megabytes for the calculation is         100
          7    Helium-argon-xenon mixture
  The gas properties are:-
    The stream number density is  1.406740000000000E+021
    The stream temperature is   300.000000000000
    The stream velocity in the x direction is  0.000000000000000E+000
    The stream velocity in the y direction is  0.000000000000000E+000
    The fraction of species        1  is  0.333300000000000
    The fraction of species        2  is  0.333300000000000
    The fraction of species        3  is  0.333400000000000
Cylindrical flow
The minimum x coordinate is   0.000000000000000E+000
The minimum x coordinate is an axis or center
The maximum x coordinate is   0.100000000000000
The maximum x coordinate is a solid surface
The maximum x boundary is a surface with the following properties
      The temperature of the surface is   300.000000000000
      The fraction of specular reflection at this surface is
  0.000000000000000E+000
      The velocity in the y direction of this surface is   500.000000000000
There are no radial weighting factors
The flowfield is initially the stream(s) or reference gas
There is no secondary stream initially at x > 0
The desired number of molecules in a sampling cell is         5000
```

The x coordinate is in the radial direction and the z coordinate is the axis of the cylinder. The y coordinate is therefore in the circumferential direction and the rotating flow is produced by the velocity of the surface in this direction. The gas is in the laboratory frame of reference so that the molecules move in straight lines.

The steady state of the flow is resembles a "solid body" rotation with circumferential velocity proportion to the radius. Figure 7.54 shows a plot of the circumferential velocity. The slip velocity at the surface was a small fraction of 1 m/s and could not be distinguished from the statistical scatter. The net shear stress was similarly small and the difference between the incident and reflected heat flux was less than one part in 100,000. The temperature was isothermal to within the statistical scatter and there was no measurable temperature slip. Approximate theories for slip are generally related to the Knudsen numbers based on the scale length of the relevant physical quantity. It is therefore not surprising that there is no temperature slip, but the absence of velocity slip was unexpected.

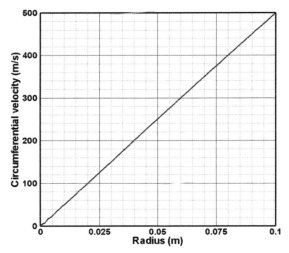

Fig. 7.54 The velocity profile.

The centrifugal effect leads to the pressure gradient that is shown in Fig. 7.55. The pressure at the surface of the cylinder almost an order of magnitude higher than the pressure at the axis and the pressure gradient at the surface is more than two orders of magnitude higher than that at the axis.

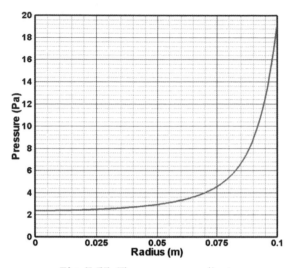

Fig. 7.55 The pressure gradient.

The pressure diffusion associated with the pressure gradient leads to the drastic species separation that is shown in Fig. 7.56.   The fraction of xenon at the axis approaches zero while the fraction at the surface doubles.   Helium drops to 10% at the surface, but more than doubles at the axis.

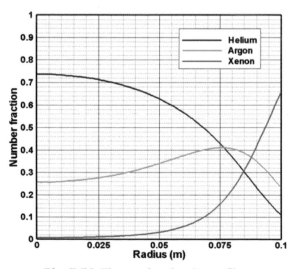

Fig. 7.56 The number density profiles.

The results reported in §12.14 of Bird (1994) are qualitatively different and correspond to an early stage of the development of the steady flow. The earlier example had a Knudsen number about one sixth that of the example in this section. The computation time to reach steady flow in this example with one million simulated molecules was about eight hours. The earlier example would require at least two days on the current computer. Even though the number of simulated molecules in the 1994 calculation would have been far fewer, the required computation time on the computers of that era would have been prohibitive. The 1994 calculation has been repeated and this has confirmed that, for a sufficiently long computation, the results are qualitatively similar to those from the test case in this section.

References

Bird, G. A. (1994). *Molecular Gas Dynamics and the Direct Simulation of Gas Flows,* Clarendon Press, Oxford.

Chapman, S. and Cowling, T. G. (1970). *The Mathematical Theory of Non-uniform Gases* 3rd edn., Cambridge University Press.

Lordi, J. A. and Mates, R. E. (1970). Rotational relaxation in nonpolar diatomic gases. *Phys. Fluids* **13,** 291-308.

Parker, J. G. (1959). Rotational and vibrational relaxation in diatomic gases. *Phys. Fluids* **2,** 449-462.

8

TWO-DIMENSIONAL FLOW APPLICATIONS

8.1 Temporary use of the DS2V program

The **DSMC** program is to be extended to deal with two and three-dimensional flows. The additional code will be based on that in the existing **DS2V** and **DS3V** programs but there are aspects of the older codes that make this a demanding task.

(i) The **DS2V/3V** codes have evolved from earlier codes over several decades and, instead of being clean codes with a logical structure, many newer procedures are clumsy modifications of older procedures. There are redundant variables and statements that confuse the logic.

(ii) The floating point variables in the older codes are almost all 32 bit and the arithmetic is accurate to only five or six significant figures. This has necessitated a large number of statements that attempt to deal with round-off error. This particularly affects the marking of elements as being inside or outside the flow and makes **DS3V** almost unusable for some problems. Many molecules have to be removed from the flow because round-off has allowed them to improperly penetrate surfaces and boundaries. In addition, the continual rounding of intermediate results means that the values of variables are subject to random walks that limit the overall accuracy of the results. A consistent 64 bit code should eliminate these problems.

(iii) Many variables in the older codes are not defined by comment statements and the codes are unnecessarily difficult to work with.

A start has been made with the extension of the **DSMC** code but, because it is not possible to predict the duration of the effort, the two-dimensional examples have been calculated with the **DS2V** program. **DS2V** has been widely used and can be regarded as being of "commercial quality". The latest procedures for chemical reactions have not been implemented in **DS2V**. The following examples will serve as test cases for the extended **DSMC** code that will appear with the next version of this book. Because of the interim nature of the current work, the source codes **DS2.F90** and **DS2V.RBC** are, like **DSMC.F90** and its auxiliary programs, being made freely available.

Descriptions of and links to the programs are in Appendix B

213

8.2 The demonstration case

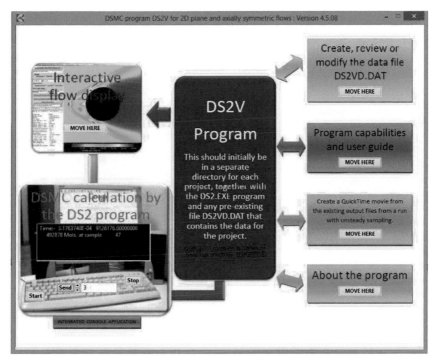

Fig. 8.1 The introductory screen of the DS2V program.

The **DS2V** program is a graphical shell program that provides menus for the generation of data files, allows the interactive running of the **DS2.F90** program and displays all result during and/or after run. It has been supplied with a demonstration data file for the Mach 5 flow of air past a sphere with an upstream pointing jet of nitrogen. This data file **DS2VD.DAT** is best reviewed by selecting the data screen, but this contains a number of tabbed menus such that the complete data cannot be shown in a single screen image. Instead, the **DS2VD.TXT** file that is generated when the **DS2** program is started lists the data as:

```
The n in version number n.m is      4
The m in version number n.m is      5
The approximate number of megabytes for the calculation is      84
The flow is axially symmetric
```

There are radial weighting factors
x limits of flowfield are -0.5000000   , 0.5000000
y limits of flowfield are 0.0000000E+00 , 0.5000000
The approximate fraction of the bounding rectangle occupied by flow is
  0.8000000
The estimated ratio of the average number density to the reference value is
  1.000000
The number of molecular species is   2
Maximum number of vibrational modes of any species is   0
The number of chemical reactions is   0
The number of surface reactions is   0
The reference diameter of species   1 is 4.0699999E-10
The reference temperature of species   1 is 273.0000
The viscosity-temperature power law of species   1 is 0.7700000
The reciprocal of the VSS scattering parameter of species 1 is   1.000000
The molecular mass of species   1 is 5.3120001E-26
Species   1 is described as Oxygen
Species   1 has electrical charge of   0
Species   1 has   2 rotational degrees of freedom
and the constant relaxation collision number is 5.000000
The reference diameter of species   2 is 4.1699999E-10
The reference temperature of species   2 is 273.0000
The viscosity-temperature power law of species   2 is 0.7400000
The reciprocal of the VSS scattering parameter of species 2 is   1.000000
The molecular mass of species   2 is 4.6500001E-26
Species   2 is described as Nitrogen
Species   2 has electrical charge of   0
Species   2 has   2 rotational degrees of freedom
and the constant relaxation collision number is 5.000000
The number density of the stream or reference gas is 3.0000001E+20
The stream temperature is 100.0000
The velocity component in the x direction is 1000.000
The fraction of species   1 is 0.2000000
The fraction of species   2 is 0.8000000
There are   1 separate surfaces
The number of points on surface   1 is   151
The maximum number of points on any surface is   151
The total number of solid surface groups is   2
The total number of solid surface intervals is   145
The total number of flow entry elements is   5
Surface   1 is defined by   151 points
Surface   1 is comprised of   1 segments
Segment   1 is a circular arc
The arc is in an anticlockwise direction
The arc is a circle with center at 0.0000000E+00 0.0000000E+00
The segment starts at 0.2200000   0.0000000E+00

The segment ends at -0.2200000   0.0000000E+00
The number of sampling property intervals along this segment is   150
The data on the   150 intervals is in   2 groups
Group   1 is a solid surface containing   145 intervals
The gas-surface interaction is species independent
Diffuse reflection at a temperature of   300.0000
The in-plane velocity of the surface is 0.0000000E+00
The angular velocity of the surface is 0.0000000E+00
For all molecular species
The gas-surface interaction is diffuse
Rotational energy accomm. coeff.   1.000000
The fraction of specular reflection is 0.0000000E+00
The fraction adsorbed is 0.0000000E+00
Group   2 is an entry line containing   5 intervals
The velocity component in the x direction is -1000.000
The velocity component in the Y direction is 0.0000000E+00
The velocity component in the Z direction is 0.0000000E+00
The gas temperature is   300.0000
Any vibrational temperature is   300.0000
The number density is 1.0000000E+21
The fraction of species   1 is 0.0000000E+00
The fraction of species   2 is   1.000000
The flow commences at time 0.0000000E+00
The flow ceases at time 1.0000000E+20
The interval is a plane of symmetry when the flow is stopped
The side at the minimum x coord is a stream boundary
The side at the maximum x coord is a stream boundary
The side at the minimum y coord is the axis of symmetry
The side at the maximum y coord is a stream boundary
The stream is the initial state of the flow
No molecules enter from a DSMIF.DAT file
The initial gas is homogeneous at the stream conditions
The sampling is for an eventual steady flow
The calculation employs the standard computational parameters

This is an axially-symmetric flow that employs weighting factors. There is a single open surface that has both ends on the axis. A single surface can be comprised of multiple straight line or circular arc segments. There is only one segment in this case and, because the flow is on the right hand side of the surface when looking from the first point, the arc is drawn from the back to the front. The surface is divided into 150 equal sampling intervals and the first 145 are in a group that is defined as a solid surface. The remaining five intervals are in a group defined as inflow intervals to set the upstream facing jet.

The Mach number contours are shown in Fig. 8.2 and the temperature contours in Fig. 8.3. The jet effectively acts as a blunt spike with radius much less than that of the sphere. The dividing streamline defines the effective shape of the body and, near the axis this is between back-to-back shock waves. The highest temperatures and the most extensive region of hot gas is associated with the reversal of the supersonic jet rather than with the bow shock. The bow shock is reinforced by a second shock that arises from the jet impingement on the sphere. There is a recirculation vortex between the edge of the jet and the impingement region. There is a small region of separated flow at the rear of the sphere and the flow in that region appears to be unsteady.

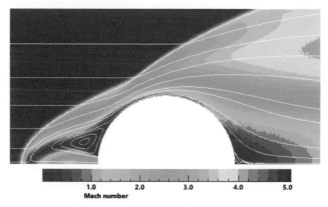

Fig. 8.2 Mach number contours.

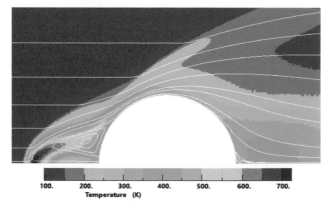

Fig. 8.3 Temperature contours.

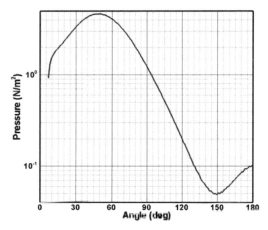

Fig. 8.4 The pressure distribution over the sphere.

The maximum pressure on the surface is in the vicinity of the jet impingement region, while the pressure adjacent to the jet is below that at 90° where the surface is parallel to the flow. Figure 8.4 also shows that the minimum pressure is at 150° and the region of positive pressure gradient is much larger than the separated flow region.

There is 20% oxygen in the "ideal air" stream and the jet is pure nitrogen. The percentage of oxygen that is shown in Fig. 8.5 therefore gives a good indication of the degree of mixing between the stream and the jet. The mixing is surprisingly rapid and this is almost certainly due to the recirculation vortex.

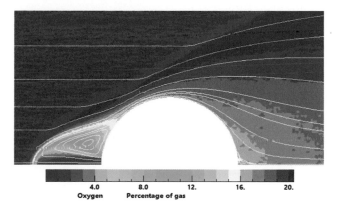

Fig. 8.5 The oxygen concentration over the flowfield.

8.3 Hypersonic flow past a cylinder

The second example is a "test case" flow for which results are available from a number of DSMC codes. It is the Mach ten flow of a monatomic gas past a cylinder at a Knudsen number of 0.0091 and a surface to freestream temperature of 2.5. The gas is argon, but with a diameter that differs slightly from the value in Appendix A and the gas database in **DSMC.f90**. The data for the calculation was reported in the **DS2VD.TXT** file as:

```
The n in version number n.m is         4
 The m in version number n.m is         5
The approximate number of megabytes for the calculation is        150
The flow is two-dimensional
x limits of flowfield are -0.2000000    ,  0.6500000
y limits of flowfield are  0.0000000E+00 ,  0.4000000
The approximate fraction of the bounding rectangle occupied by flow is
 0.9000000
The estimated ratio of the average number density to the reference value
is
   1.200000
The number of molecular species is         1
Maximum number of vibrational modes of any species is         0
The number of chemical reactions is        0
The number of surface reactions is        0
The reference diameter of species       1  is  3.5950001E-10
The reference temperature of species        1  is   1000.000
The viscosity-temperature power law of species       1  is  0.7400000
The reciprocal of the VSS scattering parameter of species       1  is
   1.000000
The molecular mass of species        1  is  6.6300000E-26
Species         1  is described as Argon
Species         1  has electrical charge of        0
Species         1  has        0 rotational degrees of freedom
The number density of the stream or reference gas is  4.2470000E+20
The stream temperature is   200.0000
The velocity component in the x direction is   2634.100
The velocity component in the y direction is  0.0000000E+00
The velocity component in the z direction is  0.0000000E+00
The fraction of species        1  is   1.000000
There are         1  separate surfaces
The number of points on surface        1  is        91
The maximum number of points on any surface is        91
The total number of solid surface groups is        1
The total number of solid surface intervals is        90
The total number of flow entry elements is        0
```

```
Surface        1  is defined by        91  points
Surface        1  is comprised of       1  segments
Segment        1  is a circular arc
The arc is in an anticlockwise direction
 The arc is a circle with center at  0.1524000      0.0000000E+00
The segment starts at  0.3048000       0.0000000E+00
The segment ends at  0.0000000E+00  0.0000000E+00
 The number of sampling property intervals along this segment is      90
 The data on the        90  intervals is in        1  groups
Group          1  is a solid surface containing      90  intervals
The gas-surface interaction is species independent
Diffuse reflection at a temperature of    500.0000
The in-plane velocity of the surface is  0.0000000E+00
The normal-to-plane velocity of the surface is  0.0000000E+00
For all molecular species
The gas-surface interaction is diffuse
Rotational energy accomm. coeff.   1.000000
The fraction of specular reflection is  0.0000000E+00
The fraction adsorbed is  0.0000000E+00
The side at the minimum x coord is
a stream boundary
The side at the maximum x coord is
a stream boundary
The side at the minimum y coord is
a plane of symmetry
The side at the maximum y coord is
a stream boundary
The stream is the initial state of the flow
No molecules enter from a DSMIF.DAT file
The initial gas is homogeneous at the stream conditions
The sampling is for an eventual steady flow
The calculation employs the standard computational parameters
```

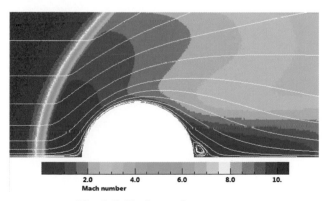

Fig. 8.6 Mach number contours.

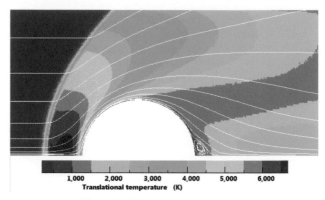

Fig. 8.7 Temperature contours.

The Mach number contours in Fig. 8.6 provide the best indication of the thickness of the bow shock. The initial Mach number contours are well upstream of the initial temperature contours in Fig. 8.7. The freestream is marked by a transition to a mottled field of the two darkest reds that mark "less than ten" and "greater than ten". This transition at the centre-line of the flow corresponds with the upstream flow boundary where there is already a temperature rise of the order of one degree. The "shock standoff distance" associated with this transition is almost double the apparent standoff distance based on the number density contours that are shown in Fig. 8.8. While the traditional classifications of flow regimes regard a flow with a Knudsen number less than 0.01 as "continuum", the continuum assumption of a discontinuous shock wave is nonsense in this case.

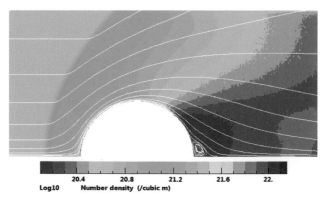

Fig. 8.8 Number density contours.

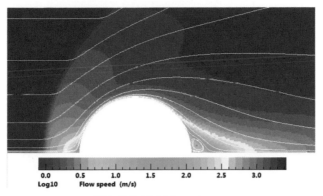

Fig. 8.9 Speed contours.

The streamlines remain straight until the temperature increase is largely complete and the flow deflection occurs with the density increase. Figure 8.9 shows that, as would be expected from the continuity equation, the speed decrease corresponds to the density increase. The number density varies by more than two orders of magnitude over the flowfield and the highest densities are confined to a very thin layer near the surface.

Separated wake flow is a common feature of hypersonic flows past cylinders and this is no exception. There is a well-defined vortex in the separated region with a height of about one quarter of the radius of the cylinder and with the stagnation point on the plane of symmetry a similar distance behind the cylinder.

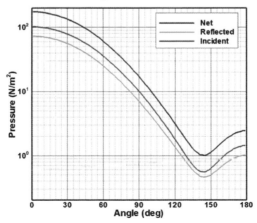

Fig. 8.10 The pressure distribution along the surface.

The surface pressure varies by more than two orders of magnitude, but the low pressure is limited by the pressure rise in the separated flow region. As well as the net pressure, Fig. 8.10 shows the separate contributions of the incident and reflected molecules. The calculation has assumed diffuse reflection with full accommodation to the surface temperature and the contributions would be equal in an equilibrium gas with no slip at the surface. The substantially greater contribution from the incident molecules is indicative of the nonequilibrium that leads to velocity and temperature slips at the surface. Equations (4.69) and (4.70) were used to directly sample the slips from the velocities of the incident and reflected molecules and the results are shown in Figs. (8.11) and (8.12).

The surface properties are sampled over two degree intervals and the sample ranges from nine million at the stagnation point to fifty five thousand at the flow separation point near 143°. The stagnation point pressure therefore has a standard deviation of only one part in three thousand and is 174.5 ± 0.1 N/m². However, there is noticeable scatter associated with the velocity and temperature slips. This is because, unlike the net pressure for which the incident and reflected components are both positive, the slips are the result of small differences in the velocity distributions of the incident and reflected molecules. While the velocity slip is zero at the stagnation points, the temperature slip is substantial over the whole surface. Both the slips have a maximum just upstream of the location of minimum pressure.

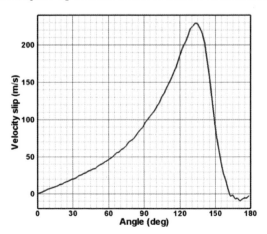

Fig. 8.11 The velocity slip at the surface.

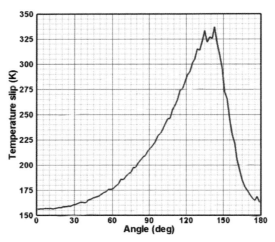

Fig. 8.12 The temperature slip at the surface.

The slip velocity becomes negative when the direction of the flow adjacent to the surface is also negative and, as expected and as shown in Fig. 8.13, the shear stress at the surface is also negative in this region.   The pressure gradient becomes positive at 143.5°, but the velocity slip and shear stress do not become negative until 163°.

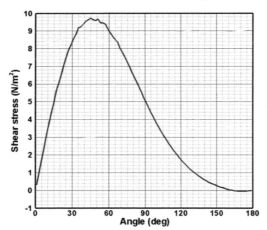

Fig. 8.13 The surface shear stress distribution.

While the larger scatter in the temperature slip plot is expected, there appears to be a persistent irregularity that is probably related to the shape of the cells near the surface.

The net heat transfer to the surface is the sum of the positive heat flux due to the incident molecules and the negative heat flux due to the reflected molecules. It is difficult to obtain an accurate result when the contributions are of similar magnitude but, in this case as shown in Fig. 8.14, the magnitude of the incident heat flux is approximately twice that of the reflected heat flux. There is therefore little scatter in the net heat transfer, but this quantity has proved to be the most sensitive to the computational parameters. The reporting of test calculations with the various DSMC codes has concentrated on the net heat transfer at the stagnation point.

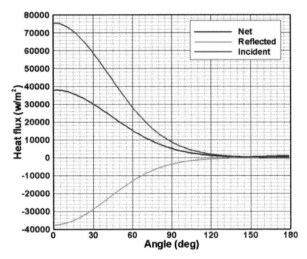

Fig. 8.14 The heat transfer distributions.

The original calculation for this test case was made by Lofthouse, Boyd and Wright (2006) and the results in the first line of Table 8.1 are from this reference. The results in the last line are from the calculation that is described in this section and the other results are from Bird (2007). The overall drag is based on a full cylinder and is double that in the current calculation that employs a semi-cylinder. The overall drag varies by almost one percent and this is many times greater than the variation in the stagnation pressure. The more sensitive heat flux at the stagnation point trends down as it converges to the correct value (Bird,2007) and the variation in the table is just over three percent.

Table 8.1 Comparative results from various DSMC codes.

Code	Molecules	Overall drag (N)	Peak heat flux (W/m²)
MONACO	26,800,000	40.00	39,319
DS2G (ver 2)	2,900,000	39.95	38,300
SMILE	24,000,000	39.76	39,000
DAC	1,300,000	39.71	38,500
DS2V	330,000	39.76	38,400
DS2V	1,007,000	39.68	38,000

The automatically generated comment in the **DS2V** output is that the calculation described in this section is "marginal" and that more molecules should be employed. The best measure of the effective cell size is the mean separation of the molecules that are chosen as collision partners. The automatic comment is based on the value of the ratio of this distance to the local mean free path, or mcs/mfp ratio. One of the three criteria for a good DSMC calculation is that this ratio should be small in comparison with unity. The second criterion is that the ratio of the time step to the mean collision time should also be small in comparison with unity and the **DS2V** program automatically sets this to a value of 0.2. The computational parameters that can optionally be set in the data allow this default value to be adjusted. The third criterion relates to the adequacy of the chosen flowfield boundaries and this relies on the (hopefully) good judgment of the user.

Fig. 8.15 The mcs/mfp distribution over the flowfield.

The mcs/mfp ratio is sampled over the flowfield and the contours are shown in Fig. 8.15. The average of this quantity is 0.103, but it exceeds 0.4 in a very thin layer adjacent to much the surface. The pressure is almost constant across the boundary layer and is not affected by the excessive mcs/mfp values. However, the heat transfer is affected and the fully converged result is just below 38,000 W/m². Because of the irregularity in the cells, there are some that differ significantly from the local average – hence the dark blue spots.

The calculations have been made with an argon diameter that was set by Lofthouse *et al* (2006). This differs from that in the built-in databases of both the **DS2V** and **DSMC** codes and it was repeated with the argon model that is used for the other calculations in the book. This is a diameter 4.17×10^{-10} m at a reference temperature of 273 K and a viscosity-temperature index of 0.81. A comparison of Figs. (8.14) and (8.16) shows that this leads to an increase in the net stagnation point heat transfer from 38,000 W/m² to 42,400 W/m². The increase is in the incident flux because the reflected heat flux, which reflects the number flux to the surface, is unchanged. The stagnation point pressure is also unchanged at 174.5 N/m², but the drag increases from 39.68 N to 40.14 N. This is because the shear stress contributes just over 8% of the drag and the shear stress with this argon model is 10% higher than that with the default model.

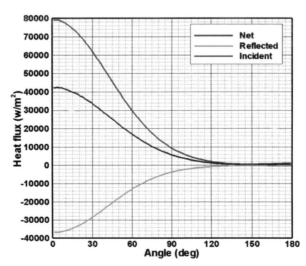

Fig. 8.16 The heat transfer distributions with the Appendix A argon model.

The effects of the change in the molecular properties from those used in the original MONACO calculation to the default values in Appendix A are surprisingly large. It is instructive to compare these effects with the errors that occur from the use of an inadequate number of simulated molecules. The calculation was therefore repeated for a range of molecule number. These calculations also provide detailed information on the way in which the various flow properties converge to the correct value as the number of simulated molecules increases.

It is the output of the mcs/mfp ratio that makes the program largely self-validating. As noted above, the program reports both the mean and maximum values of this ratio, as well as the distribution over the flowfield that is plotted in Fig. 8.15. The larger values occur in the boundary layer and are relevant to the shear stress and heat transfer, but there are generally hundreds of thousands of collision cells and the maximum value of mcs/mfp is generally associated with a rogue cell that may be anywhere in the flow. For example, while the maximum value in the above test case was reported as 0.57, the range in Fig. 8.15 has been restricted to 0.4 and, without this restriction, the red coloured region would disappear. The convergence study has therefore been based on the mean mcs/mfp ratio and the relation between this and the total number of simulated molecules is shown in Fig. 8.17.

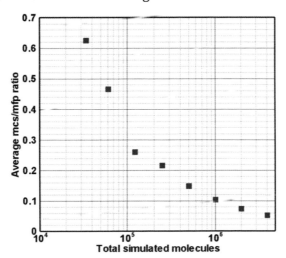

Fig. 8.17 The relationship between mcs/mfp and the molecule number.

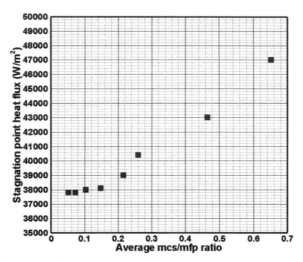

Fig. 8.18 The convergence of the stagnation point heat flux.

The convergence behavior of the stagnation point heat flux in Fig. 8.18 and that of the maximum value of the shear stress in Fig 8.19 are similar. In percentage terms, the shear stress increase with mcs/mfp is the greater but, in both cases, the rate of increase is unusually large. This is presumably due to the high values of the mcs/mfp ratio near the surface in this flow.

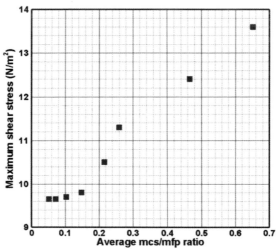

Fig. 8.19 The convergence of the maximum shear stress.

In contrast to the heat transfer and shear stress that are largely set within the boundary layer, the number flux and pressure depend primarily on the overall flow field. These quantities are listed in Table 8.2 and, while there is a slight reduction in the number flux to the converged value, there is no significant change in the pressure.

Table 8.2 Number flux and pressure at the forward stagnation point.

mcs/mfp	Number flux (10^{24}/s)	Pressure (N/m^2)
0.052	2.72	174.2
0.073	2.72	174.6
0.104	2.71	174.4
0.149	2.71	174.4
0.216	2.70	174.3
0.260	2.69	174.1
0.466	2.65	174.0
0.652	2.62	174.4

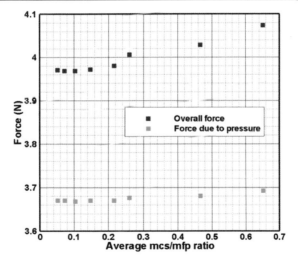

Fig. 8.20 The overall force on the full cylinder.

The result for overall force on the cylinder is significant to four digits and provides a better indication of the effect of mcs/mfp on the pressure. Fig. 8.20 shows that the drag due to pressure increases by just over 0.5%. Because of the shear contribution, there is a much larger percentage increase in the overall drag force on the cylinder.

8.4 Channel flow with specified end pressures

One of the options in DS2V is for "constant pressure boundaries". This is to allow computations for the flow through a tube or channel when the overall pressure ratio is known, but there is no information on the flow speeds. The "secondary stream" option is used to set a stationary gas at the higher pressure up to the mid-point of the channel and a stationary gas at the lower pressure in the remainder of the channel. "Reservoirs" are established at both ends of the channel. These expand to flow cross-sections that are large in comparison to that of the channel so that the flow speeds at the entry and exit are very small. The number flux of each molecular species is sampled at the mid-point and the versions of DS2V before version 4.5.10 used this flux to set the number of molecules that enter the upstream and downstream boundaries.

The first test case is for a large 5:1 pressure gradient and the data report in DS2VD.TXT is:

```
The n in version number n.m is          4
The m in version number n.m is          5
The approximate number of megabytes for the calculation is       100
The flow is two-dimensional
x limits of flowfield are -7.9999998E-04 ,  3.5999999E-03
y limits of flowfield are  0.0000000E+00 ,  2.0010001E-03
The  approximate  fraction  of  bounding  rectangle  occupied  by  flow  is
0.2000000
The  estimated  ratio  of  the  average  number  density  to  the  reference  value
is    1.000000
The number of molecular species is          1
Maximum number of vibrational modes of any species is          0
The number of chemical reactions is          0
The number of surface reactions is          0
The reference diameter of species          1  is  4.1699999E-10
The reference temperature of species          1  is   273.0000
The viscosity-temperature power law of species          1  is  0.7400000
The reciprocal of the VSS scattering parameter of species 1  is  1.000000
The molecular mass of species          1  is  4.6500001E-26
Species          1  is described as Nitrogen
Species          1  has electrical charge of          0
Species          1  has          2  rotational degrees of freedom
and the constant relaxation collision number is   5.000000
The number density of the stream or reference gas is  2.5000000E+23
The stream temperature is   300.0000
The velocity component in the x direction is  0.0000000E+00
The velocity component in the y direction is  0.0000000E+00
```

The velocity component in the z direction is 0.0000000E+00
The fraction of species   1 is   1.000000
There are   1 separate surfaces
The number of points on surface   1 is   99
The maximum number of points on any surface is   99
The total number of solid surface groups is   1
The total number of solid surface intervals is   98
The total number of flow entry elements is   0
Surface   1 is defined by   99 points
Surface   1 is comprised of   3 segments
Segment   1 is a straight line
The segment starts at -7.9999998E-04 1.0000000E-03
The segment ends at 0.0000000E+00 1.9999999E-04
 The number of sampling property intervals along segment is   16
Segment   2 is a straight line
The segment ends at 2.0000001E-03 1.9999999E-04
 The number of sampling property intervals along segment is   50
Segment   3 is a straight line
The segment ends at 3.5999999E-03 2.0000001E-03
 The number of sampling property intervals along segment is   32
 The data on the   98 intervals is in   1 groups
Group   1 is a solid surface containing   98 intervals
The gas-surface interaction is species independent
Diffuse reflection at a temperature of   300.0000
The in-plane velocity of the surface is 0.0000000E+00
The normal-to-plane velocity of the surface is 0.0000000E+00
For all molecular species
The gas-surface interaction is diffuse
Rotational energy accomm. coeff.   1.000000
The fraction of specular reflection is 0.0000000E+00
The fraction adsorbed is 0.0000000E+00
The side at the minimum x coord is a constant pressure boundary
The side at the maximum x coord is a constant pressure boundary
The side at the minimum y coord is a plane of symmetry
The side at the maximum y coord is not in the flow
The stream is the initial state of the flow
No molecules enter from a DSMIF.DAT file
A flow separation boundary divides the stream and secondary stream at
x = 1.0000000E-03
The number density of the secondary stream is 4.9999999E+22
The secondary stream temperature is   300.0000
The velocity component in the x direction is 0.0000000E+00
The velocity component in the y direction is 0.0000000E+00
The fraction of species   1 in the secondary stream is   1.000000
The sampling is for an eventual steady flow
The calculation employs the standard computational parameters

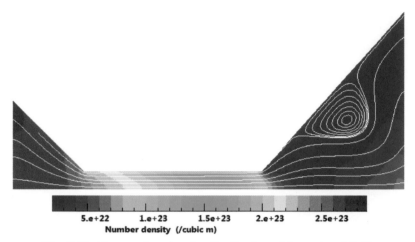

Fig. 8.21 The number density distribution with version 4.5.09.

A steady flow was established with a number flux of 5.65×10^{21} molecules per second and the number density distribution shown in Fig. 8.21. However, when the pressure ratio was reduced from 5:1 to values of the order of 1.2:1 there was an indefinite increase on the mass flux with time. This was because a positive fluctuation in the number flux led to an increase in the number of entering molecules that increased the entry pressure and led to a further increase in the number flux. The system was therefore unstable and steady flow was attained in the high pressure ratio case only because the flow was choked with a region of supersonic flow downstream of the channel.

The program was modified such that the molecule entry flux was reduced in direct proportion to the increase in pressure near the entry to the upstream reservoir. However, this only slowed the increase in mass flux and, while this would eventually be reversed, the magnitude of the oscillations would be unacceptable. The solution was to enforce steady flow by keeping the total number of simulated molecules fixed to the initial value. This was achieved by setting the number of molecules that entered the downstream reservoir to the number leaving that boundary less the net number change at the upstream boundary. Also, because there had been little change in the molecule number in the earlier calculation and the average entry speed was about 25 m/s, the data was altered such that the velocity component in the x direction of the main stream was 25 m/s rather than zero.

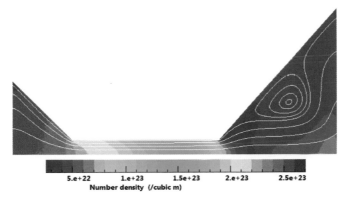

Fig. 8.22 The number density distribution with version 4.5.10.

The new boundary conditions lead to little qualitative change but, as shown in Fig. 8.22, the number density in the upstream reservoir is slightly lower and the number flux falls to 5.31×10^{21} molecules per second. The Mach number contours in Fig. 8.23 show the choked flow with largely sonic flow at the exit of the channel. The high speed flow from the tube is unable to cope with the rapid expansion and the flow in the downstream reservoir is dominated by a large vortex. The Knudsen number based on the upstream mean free path and the full height of the channel is 0.0125. This is near continuum, but the boundary layer has a significant effect and the number flux is less than half the inviscid continuum number flux of 1.22×10^{22} molecules per second. The degree of influence of the reservoir shape on the number flux would be an interesting study.

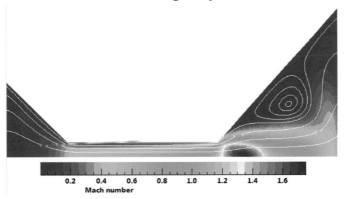

Fig. 8.23 The Mach number distribution.

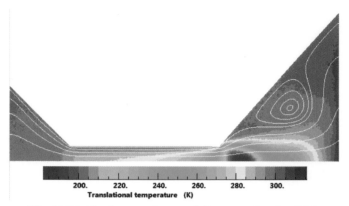

Fig. 8.24 The temperature variation over the flowfield.

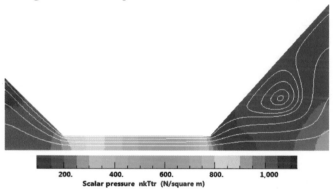

Fig. 8.25 The scalar pressure contours.

Figures 8.24 and 8.25 show the temperature and scalar pressure contours. The latter quantity involves the product of the number density and temperature. The large temperature variations are characteristic of flows in which the velocities are at least of the order of the speed of sound. The statistical fluctuations in the velocities are relative to the average molecular speed and this speed differs from the speed of sound by a numerical factor near unity. DSMC is therefore more suited to large disturbance flows and it is frequently claimed that DSMC is not capable of dealing with typical MEMS applications that involve flow speeds that are small in comparison with the speed of sound.

To test these claims, the calculation was repeated with the overall pressure and number density difference reduced by a factor of ten. That is a downstream number density of 2.3×10^{23} /m^3.

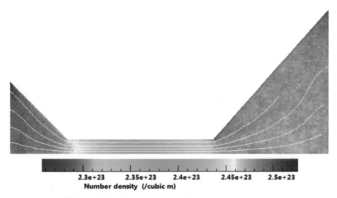

Fig. 8.26 The number density contours.

The number density is shown in Fig. 8.26 and the speed distribution over the flowfield in Fig. 8.27. The number flux was 1.27×10^{21} molecules per second and, at the mid-point of the channel where the number density is 2.39×10^{23} /m^3, the average speed is 26.6 m/s. This was established to reasonable accuracy through a short preliminary calculation and, because the area of the upstream reservoir entry is five times that of the channel, the stream was assigned a velocity component of 5 m/s in the x direction.

The results are based on a run that lasted two days on a single processor. There is therefore no problem in the application of the DSMC method to problems with flow speeds of the order of 10 m/s. Problems with velocities less than 1 m/s require a sample of the order of 10^8. It is shown in §8.9 that this can now be attained, even on a personal computer, and claims that DSMC cannot be applied to MEMS type problems are now outdated.

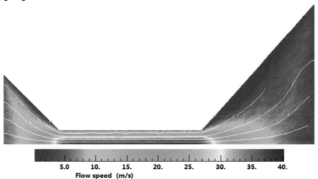

Fig. 8.27 The flow speed contours.

8.5 Reacting flow over a re-entry vehicle

This example is for a 70° blunt-cone heat shield with a diameter of 2 m that has a cylindrical after-body of diameter 1.4 m moving at 7,500 m/s in air at a number density of 1×10^{20} m/s. The corresponding altitude is just under 100 km where the temperature is assumed to be 200 K. The Mach number is 26.4 and the Knudsen number based on the diameter of the heat shield is 0.01. The heat shield is assumed to be at a temperature of 1,000 K and the after-body at a temperature of 300 K. The chemical reactions employ the TCE model because the Q-K option in **DS2V** is an early and incomplete implementation that should not be selected. There is one surface reaction in that the heat shield is 2% catalytic to atomic oxygen.

The data is reported in DS2VD.TXT as:

```
Data summary for program DS2V
 The n in version number n.m is          4
 The m in version number n.m is          5
 The approximate number of megabytes for the calculation is          200
 The flow is axially symmetric
 There are radial weighting factors
 x limits of flowfield are -0.3000000     ,   1.500000
 y limits of flowfield are  0.0000000E+00 ,   1.500000
 The approx. fraction of the bounding rectangle occupied by flow is  0.7000
 The estimated ratio of the average number density to ref. value is   1.000
 The number of molecular species is     5
 Maximum number of vibrational modes of any species is          1
 The number of chemical reactions is       23
 The number of third-body tables is        4
 The number of surface reactions is        1
 The reference diameter of species        1  is  4.0699999E-10
 The reference temperature of species     1  is   273.0000
 The viscosity-temperature power law of species     1  is  0.7700000
 The reciprocal of the VSS scattering parameter of species 1  is   1.000000
 The molecular mass of species           1  is  5.3120001E-26
 Species          1  is described as Oxygen
 Species          1  has          2  rotational degrees of freedom
 and the constant relaxation collision number is   5.000000
 Species          1  has          1  vibrational modes
 Char. temperature of mode     1  of species       1  is   2256.000
 The reference value of the vibrational collision number is    56.50000
 This is at a reference temperature of    153.5000
 The reference diameter of species        2  is  4.1699999E-10
 The reference temperature of species     2  is   273.0000
 The viscosity-temperature power law of species     2  is  0.7400000
 The reciprocal of the VSS scattering parameter of species 2  is   1.000000
 The molecular mass of species           2  is  4.6500001E-26
 Species          2  is described as Nitrogen
 Species          2  has          2  rotational degrees of freedom
 and the constant relaxation collision number is   5.000000
```

```
Species          2   has             1   vibrational modes
Char. temperature of mode  1  of species         2  is   3371.000
The reference value of the vibrational collision number is   9.100000
This is at a reference temperature of    220.0000
The reference diameter of species          3  is  3.0000000F-10
The reference temperature of species       3  is   273.0000
The viscosity-temperature power law of species        3  is  0.8000000
The reciprocal of the VSS scattering parameter of species 3  is   1.000000
The molecular mass of species          3  is  2.6560001E-26
Species          3  is described as Atomic_oxygen
Species          3  has             0  rotational degrees of freedom
The reference diameter of species          4  is  3.0000000E-10
The reference temperature of species       4  is   273.0000
The viscosity-temperature power law of species        4  is  0.8000000
The reciprocal of the VSS scattering parameter of species 4  is   1.000000
The molecular mass of species          4  is  2.3250000E-26
Species          4  is described as Atomic_nitrogen
Species          4  has             0  rotational degrees of freedom
The reference diameter of species          5  is  4.2000001E-10
The reference temperature of species       5  is   273.0000
The viscosity-temperature power law of species        5  is  0.7900000
The reciprocal of the VSS scattering parameter of species 5  is   1.000000
The molecular mass of species          5  is  4.9800000E-26
Species          5  is described as Nitric oxide
Species          5  has             2  rotational degrees of freedom
and the constant relaxation collision number is   5.000000
Species          5  has             1  vibrational modes
Char. temperature of mode  1  of species         5  is   2719.000
The constant vibrational collision number is   50.00000
 The following data is for gas phase reaction              1
Species code of one molecule is           1
Species code of the other molecule is          4
First post-reaction species code in the dissociation          3
Second post-reaction species code          3
Third post-reaction species code          4
The number of contributing internal degrees of freedom   1.000000
The activation energy  8.1970000E-19
1.E10 times the pre-exponential parameter  5.9930000E-02
The temperature exponent in the rate equation   -1.000000
The energy of the reaction -8.1970000E-19
 The following data is for gas phase reaction              2
Species code of one molecule is           1
Species code of the other molecule is          5
First post-reaction species code in the dissociation          3
Second post-reaction species code          3
Third post-reaction species code          5
The number of contributing internal degrees of freedom   1.000000
The activation energy  8.1970000E-19
1.E10 times the pre-exponential parameter  5.9930000E-02
The temperature exponent in the rate equation   -1.000000
The energy of the reaction -8.1970000E-19
The following data is for gas phase reaction              3
 Species code of one molecule is           1
Species code of the other molecule is          2
First post-reaction species code in the dissociation          3
```

```
Second post-reaction species code           3
Third post-reaction species code          2
The number of contributing internal degrees of freedom   1.500000
The activation energy  8.1970000E-19
1.E10 times the pre-exponential parameter  0.1198000
The temperature exponent in the rate equation  -1.000000
The energy of the reaction -8.1970000E-19
The following data is for gas phase reaction          4
Species code of one molecule is           1
Species code of the other molecule is           1
First post-reaction species code in the dissociation          3
Second post-reaction species code           3
Third post-reaction species code          1
The number of contributing internal degrees of freedom   1.500000
The activation energy  8.1970000E-19
1.E10 times the pre-exponential parameter  0.5393000
The temperature exponent in the rate equation  -1.000000
The energy of the reaction -8.1970000E-19
The following data is for gas phase reaction          5
Species code of one molecule is           1
Species code of the other molecule is          3
First post-reaction species code in the dissociation          3
Second post-reaction species code          3
Third post-reaction species code          3
The number of contributing internal degrees of freedom   1.000000
The activation energy  8.1970000E-19
1.E10 times the pre-exponential parameter   1.498000
The temperature exponent in the rate equation  -1.000000
The energy of the reaction -8.1970000E-19
The following data is for gas phase reaction          6
Species code of one molecule is           2
Species code of the other molecule is          3
First post-reaction species code in the dissociation          4
Second post-reaction species code          4
Third post-reaction species code          3
The number of contributing internal degrees of freedom  0.5000000
The activation energy  1.5610000E-18
1.E10 times the pre-exponential parameter  3.1870001E-03
The temperature exponent in the rate equation -0.5000000
The energy of the reaction -1.5610000E-18
The following data is for gas phase reaction          7
Species code of one molecule is           2
Species code of the other molecule is          1
First post-reaction species code in the dissociation          4
Second post-reaction species code          4
Third post-reaction species code          1
The number of contributing internal degrees of freedom  0.5000000
The activation energy  1.5610000E-18
1.E10 times the pre-exponential parameter  3.1870001E-03
The temperature exponent in the rate equation -0.5000000
The energy of the reaction -1.5610000E-18
The following data is for gas phase reaction          8
Species code of one molecule is           2
Species code of the other molecule is          5
First post-reaction species code in the dissociation          4
```

Second post-reaction species code   4
Third post-reaction species code   5
The number of contributing internal degrees of freedom 0.5000000
The activation energy 1.5610000E-18
1.E10 times the pre-exponential parameter 3.1870001E-03
The temperature exponent in the rate equation -0.5000000
The energy of the reaction -1.5610000E-18
The following data is for gas phase reaction   9
Species code of one molecule is   2
Species code of the other molecule is   2
First post-reaction species code in the dissociation   4
Second post-reaction species code   4
Third post-reaction species code   2
The number of contributing internal degrees of freedom 1.000000
The activation energy 1.5610000E-18
1.E10 times the pre-exponential parameter 7.9680001E-03
The temperature exponent in the rate equation -0.5000000
The energy of the reaction -1.5610000E-18
The following data is for gas phase reaction   10
Species code of one molecule is   2
Species code of the other molecule is   4
First post-reaction species code in the dissociation   4
Second post-reaction species code   4
Third post-reaction species code   4
The number of contributing internal degrees of freedom 1.000000
The activation energy 1.5610000E-18
1.E10 times the pre-exponential parameter   690.0000
The temperature exponent in the rate equation -1.500000
The energy of the reaction -1.5610000E-18
The following data is for gas phase reaction   11
Species code of one molecule is   5
Species code of the other molecule is   2
First post-reaction species code in the dissociation   4
Second post-reaction species code   3
Third post-reaction species code   2
The number of contributing internal degrees of freedom 1.000000
The activation energy 1.0430000E-18
1.E10 times the pre-exponential parameter   6.590000
The temperature exponent in the rate equation -1.500000
The energy of the reaction -1.0430000E-18
The following data is for gas phase reaction   12
Species code of one molecule is   5
Species code of the other molecule is   1
First post-reaction species code in the dissociation   4
Second post-reaction species code   3
Third post-reaction species code   1
The number of contributing internal degrees of freedom 1.000000
The activation energy 1.0430000E-18
1.E10 times the pre-exponential parameter   6.590000
The temperature exponent in the rate equation -1.500000
The energy of the reaction -1.0430000E-18
The following data is for gas phase reaction   13
Species code of one molecule is   5
Species code of the other molecule is   5
First post-reaction species code in the dissociation   4

```
Second post-reaction species code           3
Third post-reaction species code            5
The number of contributing internal degrees of freedom   1.000000
The activation energy  1.0430000E-18
1.E10 times the pre-exponential parameter   131.8000
The temperature exponent in the rate equation  -1.500000
The energy of the reaction -1.0430000E-18
The following data is for gas phase reaction          14
Species code of one molecule is             5
Species code of the other molecule is           3
First post-reaction species code in the dissociation         4
Second post-reaction species code           3
Third post-reaction species code            3
The number of contributing internal degrees of freedom   1.000000
The activation energy  1.0430000E-18
1.E10 times the pre-exponential parameter   131.8000
The temperature exponent in the rate equation  -1.500000
The energy of the reaction -1.0430000E-18
The following data is for gas phase reaction          15
Species code of one molecule is             5
Species code of the other molecule is           4
First post-reaction species code in the dissociation         4
Second post-reaction species code           3
Third post-reaction species code            4
The number of contributing internal degrees of freedom   1.000000
The activation energy  1.0430000E-18
1.E10 times the pre-exponential parameter   131.8000
The temperature exponent in the rate equation  -1.500000
The energy of the reaction -1.0430000E-18
The following data is for gas phase reaction          16
Species code of one molecule is             5
Species code of the other molecule is           3
The reaction results in two post-collision particles
One post-reaction species code              1
Other post-collision species code           4
The number of contributing internal degrees of freedom  0.0000000E+00
The activation energy  2.7190001E-19
1.E10 times the pre-exponential parameter  5.2790002E-11
The temperature exponent in the rate equation   1.000000
The energy of the reaction -2.7190001E-19
The following data is for gas phase reaction          17
Species code of one molecule is             2
Species code of the other molecule is           3
The reaction results in two post-collision particles
One post-reaction species code              5
Other post-collision species code           4
The number of contributing internal degrees of freedom  0.0000000E+00
The activation energy  5.1750001E-19
1.E10 times the pre-exponential parameter  1.1200000E-06
The temperature exponent in the rate equation  0.0000000E+00
The energy of the reaction -5.1750001E-19
The following data is for gas phase reaction          18
Species code of one molecule is             1
Species code of the other molecule is           4
The reaction results in two post-collision particles
```

```
One post-reaction species code          5
Other post-collision species code          3
The number of contributing internal degrees of freedom  0.0000000E+00
The activation energy  4.9679999E-20
1.E10 times the pre-exponential parameter  1.5980000F-08
The temperature exponent in the rate equation  0.5000000
The energy of the reaction  2.7190001E-19
The following data is for gas phase reaction          19
Species code of one molecule is          5
Species code of the other molecule is          4
The reaction results in two post-collision particles
One post-reaction species code          2
Other post-collision species code          3
The number of contributing internal degrees of freedom  0.0000000E+00
The activation energy  0.0000000E+00
1.E10 times the pre-exponential parameter  2.4900001E-07
The temperature exponent in the rate equation  0.0000000E+00
The energy of the reaction  5.1750001E-19
The following data is for gas phase reaction          20
Species code of one molecule is          3
Species code of the other molecule is          3
The reaction is a recombination or an
The post-recombination species code is          1
The code of the third-body table is          1
The number of contributing internal degrees of freedom  0.0000000E+00
The activation energy  0.0000000E+00
1.E10 times the pre-exponential parameter  8.2970002E-35
The temperature exponent in the rate equation -0.5000000
The energy of the reaction  8.1970000E-19
The following data is for gas phase reaction          21
Species code of one molecule is          4
Species code of the other molecule is          4
The reaction is a recombination or an
The post-recombination species code is          2
The code of the third-body table is          2
The number of contributing internal degrees of freedom  0.0000000E+00
The activation energy  0.0000000E+00
1.E10 times the pre-exponential parameter  3.0051001E-34
The temperature exponent in the rate equation -0.5000000
The energy of the reaction  1.5610000E-18
The following data is for gas phase reaction          22
Species code of one molecule is          4
Species code of the other molecule is          4
The reaction is a recombination or an
The post-recombination species code is          2
The code of the third-body table is          3
The number of contributing internal degrees of freedom  0.0000000E+00
The activation energy  0.0000000E+00
1.E10 times the pre-exponential parameter  6.3961999E-30
The temperature exponent in the rate equation  -1.500000
The energy of the reaction  1.5637000E-18
The following data is for gas phase reaction          23
Species code of one molecule is          4
Species code of the other molecule is          3
The reaction is a recombination or an
```

```
The post-recombination species code is            5
The code of the third-body table is         4
The number of contributing internal degrees of freedom  0.0000000E+00
The activation energy  0.0000000E+00
1.E10 times the pre-exponential parameter  2.7846000E-30
The temperature exponent in the rate equation  -1.500000
The energy of the reaction  1.0430000E-18
The following data is for third-body probability         1
That for species        1  is   9.000000
That for species        2  is   2.000000
That for species        3  is   25.00000
That for species        4  is   1.000000
That for species        5  is   1.000000
The following data is for third-body probability         2
That for species        1  is   1.000000
That for species        2  is   2.500000
That for species        3  is   1.000000
That for species        4  is   0.0000000E+00
That for species        5  is   1.000000
The following data is for third-body probability         3
That for species        1  is   0.0000000E+00
That for species        2  is   0.0000000E+00
That for species        3  is   0.0000000E+00
That for species        4  is   1.000000
That for species        5  is   0.0000000E+00
The following data is for third-body probability         4
That for species        1  is   1.000000
That for species        2  is   1.000000
That for species        3  is   20.00000
That for species        4  is   20.00000
That for species        5  is   20.00000
The following data is for surface reaction          1
Surface reaction         1  is described as Surf_reac_1
Species code of the first incident molecule is         3
Species code of the second incident molecule is         3
Species code of the first reflected molecule is         1
The energy input to the surface is  8.1970000E-19
The number density of the stream or reference gas is  1.0000000E+20
The stream temperature is   200.0000
The stream vibrational temperature is   200.0000
The velocity component in the x direction is   7500.000
The fraction of species        1  is  0.2000000
The fraction of species        2  is  0.8000000
The fraction of species        3  is  0.0000000E+00
The fraction of species        4  is  0.0000000E+00
The fraction of species        5  is  0.0000000E+00
There are          2 separate surfaces
The number of points on surface         1  is          103
The number of points on surface         2  is          3
The maximum number of points on any surface is        103
The total number of solid surface groups is          2
The total number of solid surface intervals is        102
The total number of flow entry elements is         2
Surface          1  is defined by        103  points
Surface          1  is comprised of          6  segments
```

```
Segment          1  is a straight line
The segment starts at  0.8000000      0.0000000E+00
The segment ends at  0.8000000      0.7000000
 The number of sampling property intervals along segment is          6
Segment          2  is a straight line
The segment ends at  0.4168917      0.7000000
The number of sampling property intervals along segment is         10
Segment          3  is a straight line
The segment ends at  0.4168917      0.9500000
The number of sampling property intervals along  segment is          6
Segment          4  is a circular arc
The arc is in an anticlockwise direction
The arc is a circle with center at  0.3668917      0.9500000
The segment ends at  0.3199070      0.9671010
The number of sampling property intervals along segment is         15
Segment          5  is a straight line
The segment ends at  3.0153699E-02  0.1710000
The number of sampling property intervals along this segment is         50
Segment          6  is a circular arc in an anticlockwise direction
The arc is a circle with center at  0.5000000      0.0000000E+00
The segment ends at  0.0000000E+00  0.0000000E+00
The number of sampling property intervals along segment is         15
The data on the          102  intervals is in          2  groups
Group          1  is a solid surface containing         22  intervals
The gas-surface interaction is species independent
Diffuse reflection at a temperature of   300.0000
The in-plane velocity of the surface is  0.0000000E+00
The angular velocity of the surface is  0.0000000E+00
For all molecular species
The gas-surface interaction is diffuse
Rotational energy accomm. coeff.   1.000000
Vibrational energy accomm. coeff.   1.000000
The fraction of specular reflection is  0.0000000E+00
The fraction adsorbed is  0.0000000E+00
The probability of surface reaction          1  is  0.0000000E+00
Group          2  is a solid surface containing         80  intervals
The gas-surface interaction is species independent
Diffuse reflection at a temperature of   1000.000
The in-plane velocity of the surface is  0.0000000E+00
The angular velocity of the surface is  0.0000000E+00
For all molecular species
The gas-surface interaction is diffuse
Rotational energy accomm. coeff.   1.000000
Vibrational energy accomm. coeff.   1.000000
The fraction of specular reflection is  0.0000000E+00
The fraction adsorbed is  0.0000000E+00
The probability of surface reaction          1  is  2.0000000E-02
Surface          2  is defined by          3  points
Surface          2  is comprised of          2  segments
Segment          1  is a straight line
The segment starts at -0.3000000      0.1000000
The segment ends at -0.1000000      1.200000
The number of sampling property intervals along this segment is          1
Segment          2  is a straight line
The segment ends at  0.2000000      1.500000
```

```
The number of sampling property intervals along this segment is          1
The data on the              2 intervals is in          1 groups
Group            1 is a stream boundary containing        2 intervals
The velocity component in the x direction is   7500.000
The velocity component in the Y direction is  0.0000000E+00
The velocity component in the Z direction is  0.0000000E+00
The gas temperature is   200.0000
Any vibrational temperature is   200.0000
The number density is  1.0000000E+20
The fraction of species          1 is  0.2000000
The fraction of species          2 is  0.8000000
The fraction of species          3 is  0.0000000E+00
The fraction of species          4 is  0.0000000E+00
The fraction of species          5 is  0.0000000E+00
The side at the minimum x coord is a stream boundary
The side at the maximum x coord is a stream boundary
The side at the minimum y coord is the axis of symmetry
The side at the maximum y coord is a stream boundary
The stream is the initial state of the flow
No molecules enter from a DSMIF.DAT file
The initial gas is homogeneous at the stream conditions
The sampling is for an eventual steady flow
The calculation employs the standard computational parameters
```

The specifications of the five molecular species and the twenty three chemical reactions comprise a very large fraction of this long data report, but the complete "Real Air" gas model is loaded through a single mouse click in the data generation screen. **The length of the data report is therefore not indicative of the effort that is required to specify the data within the DS2V program.** In the newer **DSMC.F90** program, the gas models form part of the source code.

In addition, some data items that complicate this example are a legacy of the far slower computers when this case was first calculated. For example, it is no longer necessary to reduce the computational effort through the angled upstream boundary of "stream surfaces". A modern data set would simply have the upstream boundary of the bounding rectangle as a "stream boundary" and it would be desirable to have it further upstream.

The temperature and number density distributions over the flowfield are shown in Fig. 8.28. The maximum temperature in the bow shock is only several thousand degrees less than the non-reacting temperature behind a normal shock wave. The number density in the thin surface layer is two orders of magnitude above the freestream value. The number density in the wake is well below the freestream number density, but is sufficiently high for vortices to form in the regions of separated flow behind the vertical surfaces.

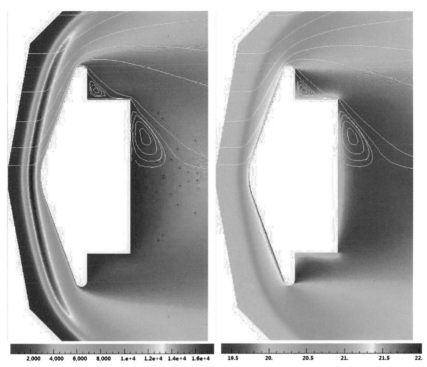

2,000 4,000 6,000 8,000 1.e+4 1.2e+4 1.4e+4 1.6e+4   19.5   20.   20.5   21.   21.5   22.

Fig. 8.28 The temperature (K) and number density ($\log_{10}$ m^{-3}) distributions.

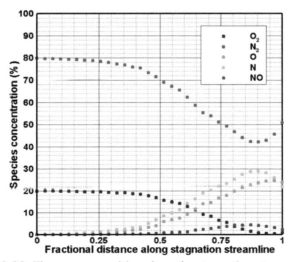

Fig. 8.29 The gas composition along the stagnation streamline.

The effect of the chemical reactions is illustrated by Fig. 8.29. Oxygen is almost completely dissociated at the stagnation point while nitrogen is about half dissociated.

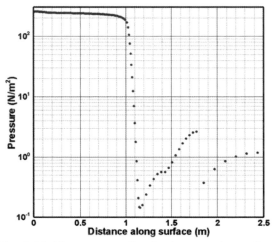

Fig. 8.30 The pressure distribution over the surface.

The overall pressure profile in Fig. 8.30 and the net heat transfer profile in Fig. 8.31 are qualitatively similar. The reattachment of the flow on the cylindrical after-body has a marked effect and the transition from a spherical to a conical surface on the heat shield has a slight effect.

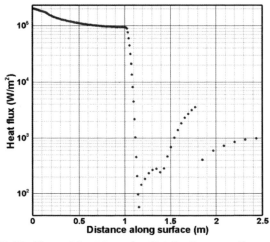

Fig. 8.31 The net heat transfer distribution over the surface.

8.6 Shock diffraction by a vertical flat plate

One of the options for the "minimum x" boundary is a specularly reflecting flat piston moving with a constant speed. The speed is set to 500 m/s in argon at 300 K and this generates a Mach 2.47 shock wave. The shock is diffracted around a vertical specularly reflecting flat plate that has a semi-height of twenty five mean free paths in the undisturbed gas. The data is reported by **DS2VD.TXT** as:

```
Data summary for program DS2V
 The n in version number n.m is          4
 The m in version number n.m is          5
 The approximate number of megabytes for the calculation is        680
 The flow is two-dimensional
 x limits of flowfield are  0.0000000E+00 ,  0.9000000
 y limits of flowfield are  0.0000000E+00 ,  0.5000000
 The approximate fraction of bounding rectangle occupied by flow is  1.0000
 The estimated ratio of the average number density to ref. value is  1.0000
 The number of molecular species is          1
 Maximum number of vibrational modes of any species is          0
 The number of chemical reactions is         0
 The number of surface reactions is         0
 The reference diameter of species          1  is  4.1699999E-10
 The reference temperature of species          1  is   273.0000
 The viscosity-temperature power law of species          1  is  0.8100000
 The reciprocal of the VSS scattering parameter of species 1  is  1.000000
 The molecular mass of species          1  is  6.6300000E-26
 Species          1  is described as Argon
 Species          1  has          0  rotational degrees of freedom
 The number density of the stream or reference gas is  1.2500000E+21
 The stream temperature is   300.0000
 The velocity component in the x direction is  0.0000000E+00
 The velocity component in the y direction is  0.0000000E+00
 The velocity component in the z direction is  0.0000000E+00
 The fraction of species          1  is  1.000000
 There are          1  separate surfaces
 The number of points on surface          1  is         65
 The maximum number of points on any surface is         65
 The total number of solid surface groups is          1
 The total number of solid surface intervals is         64
 The total number of flow entry elements is         0
 Surface          1  is defined by         65  points
 Surface          1  is comprised of          3  segments
 Segment          1  is a straight line
 The segment starts at  0.7020000       0.0000000E+00
 The segment ends at  0.7020000       0.2480000
  The number of sampling property intervals along this segment is         30
 Segment          2  is a circular arc
 The arc is in an anticlockwise direction
  The arc is a circle with center at  0.7000000       0.2480000
 The segment ends at  0.6980000       0.2480000
  The number of sampling property intervals along this segment is          4
 Segment          3  is a straight line
```

```
The segment ends at  0.6980000       0.0000000E+00
 The number of sampling property intervals along this segment is          30
 The data on the           64  intervals is in           1  groups
Group           1  is a solid surface containing          64  intervals
The gas-surface interaction is species independent
Diffuse reflection at a temperature of    300.0000
The in-plane velocity of the surface is  0.0000000E+00
The normal-to-plane velocity of the surface is  0.0000000E+00
The gas-surface interaction is diffuse
Rotational energy accomm. coeff.   1.000000
The fraction of specular reflection is    1.000000
The fraction adsorbed is  0.0000000E+00
The side at the minimum x coord is
a specular wall moving into the flow
The side at the maximum x coord is
a stream boundary
The side at the minimum y coord is
a plane of symmetry
The side at the maximum y coord is
a plane of symmetry
The speed of the moving wall is    500.0000
The wall stops moving at time    100.0000
The wall is a plane normal to the x axis
The stream is the initial state of the flow
No molecules enter from a DSMIF.DAT file
The initial gas is homogeneous at the stream conditions
The sampling is for an unsteady flow
The computational variables may be altered
The number of divisions in x & y directions is changed by factor of  1.000
The number of elements per division in the x & y directions is          10
The initial number of molecules is changed by a factor of   1.000000
The number of moves in a mean collision time is          5
The number of moves in a mean transit time is          2
The standard sampling interval is changed by a factor of   1.000000
The standard output interval is changed by a factor of  0.3000000
A time average is taken over the last 1.000000    of the sampling interval
Employ nearest neighbor collisions for all flows
```

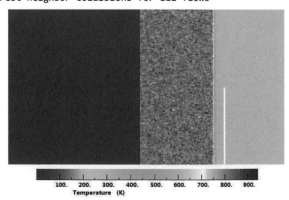

100. 200. 300. 400. 500. 600. 700. 800. 900.
Temperature (K)

Fig. 8.32 The shock wave just before its impact with the vertical flat plate.

Figure 8.32 shows the flow configuration shortly before the shock collides with the plate. The empty computational region behind the piston has been assigned zero temperature. The Rankine-Hugoniot temperature behind a Mach 2.47 shock wave in a monatomic gas at 300 K is 825.2 K.

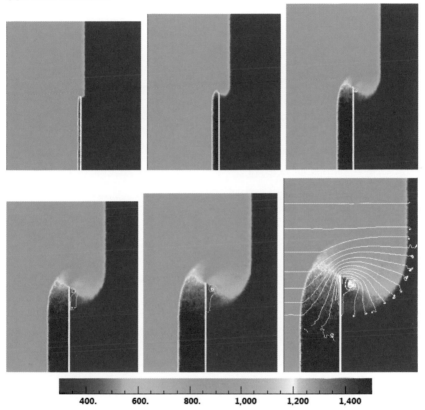

Fig. 8.33 The temperature (K) distribution at advancing times.

The temperature distributions at six advancing times after the impact of the shock with the plate are shown in Fig. 8.33, while Figs. 8.34 and 8.35 show the pressure and flow speed at the highest time. The temperature behind the reflected shock from one-dimensional theory is 1510 K and the theoretical pressures are compared with the sampled plate pressure distributions in Fig. 8.36. The most notable features are the formation of the vortex and the region of supersonic flow downstream of the edge of the plate.

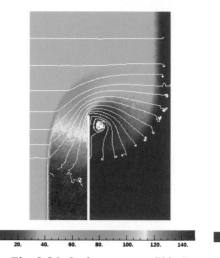

Fig. 8.34 Scalar pressure (N/m²).

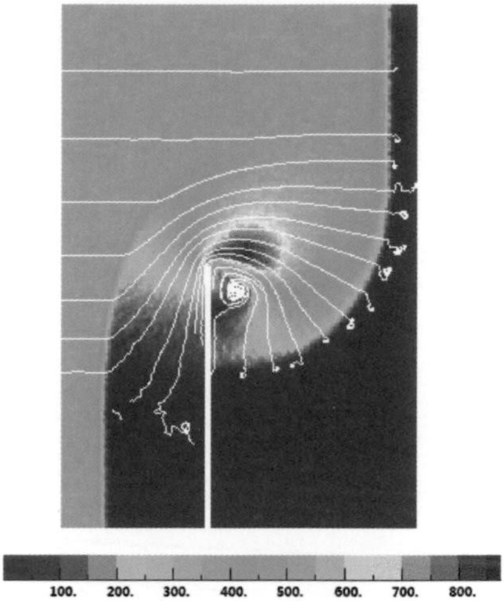

Fig. 8.35 Flow speed (m/s).

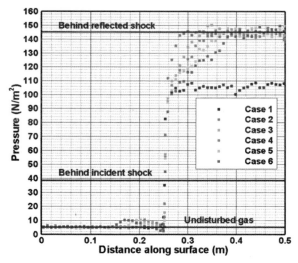

Fig. 8.36 Pressure distributions along the plate.

Because of the relatively small samples, there is significant scatter in the distributions. The shock reflection is evidently not complete at the time of the Case 1 sampling. The total computation time was less than 1.5 hours on a 2.8 GHz Intel I7 and, **for unsteady flows, the DSMC computational requirements compare well with those of typical continuum CFD calculations.**

8.7 Unstable flow past a blunt cone with ballute

This example is the Mach 5.8 flow past a blunt cone with a towed toroidal ballute. The number density of the assumed ideal air is 2×10^{20} m^{-3} and the freestream mean free path is approximately 0.005 times the total length of the combined body. The data description is:

```
Data summary for program DS2V
 The n in version number n.m is          4
 The m in version number n.m is          5
 The approximate number of megabytes for the calculation is          600
 The flow is axially symmetric
 There are radial weighting factors
 x limits of flowfield are  0.0000000E+00 ,   1.800000
 y limits of flowfield are  0.0000000E+00 ,   1.100000
 The approximate fraction of bounding rectangle occupied by flow is 0.75000
 The estimated ratio of the average number density to ref. value is 1.00000
 The number of molecular species is          2
 Maximum number of vibrational modes of any species is          0
 The number of chemical reactions is          0
 The number of surface reactions is          0
 The reference diameter of species          1  is  4.0699999E-10
 The reference temperature of species          1  is   273.0000
 The viscosity-temperature power law of species          1  is  0.7700000
 The reciprocal of the VSS scattering parameter of species 1  is   1.000000
 The molecular mass of species          1  is  5.3120001E-26
 Species          1  is described as Species_1
 Species          1  has electrical charge of          0
 Species          1  has          2  rotational degrees of freedom
 and the constant relaxation collision number is   5.000000
 The reference diameter of species          2  is  4.1699999E-10
 The reference temperature of species          2  is   273.0000
 The viscosity-temperature power law of species          2  is  0.7400000
 The reciprocal of the VSS scattering parameter of species 2  is   1.000000
 The molecular mass of species          2  is  4.6500001E-26
 Species          2  is described as Species_2
 Species          2  has electrical charge of          0
 Species          2  has          2  rotational degrees of freedom
 and the constant relaxation collision number is   5.000000
 The number density of the stream or reference gas is  2.0000000E+20
 The stream temperature is   300.0000
 The velocity component in the x direction is   2000.000
 The fraction of species          1  is  0.2000000
 The fraction of species          2  is  0.8000000
 There are          3  separate surfaces
 The number of points on surface          1  is          33
 The number of points on surface          2  is          61
 The number of points on surface          3  is           2
 The maximum number of points on any surface is          61
 The total number of solid surface groups is          2
 The total number of solid surface intervals is          92
 The total number of flow entry elements is          1
 Surface          1  is defined by          33  points
```

```
Surface          1  is comprised of          3   segments
Segment          1  is a straight line
The segment starts at  0.4000000      0.0000000E+00
The segment ends at  0.4000000      0.2309400
 The number of sampling property intervals along this segment is        6
Segment          2  is a straight line
The segment ends at  0.1500000      8.6599998E-02
 The number of sampling property intervals along this segment is       20
Segment          3  is a circular arc
The arc is in an anticlockwise direction
 The arc is a circle with center at  0.2000000       0.0000000E+00
The segment ends at  0.1000000      0.0000000E+00
 The number of sampling property intervals along this segment is        6
 The data on the        32  intervals is in        1  groups
Group          1  is a solid surface containing        32  intervals
The gas-surface interaction is species independent
Diffuse reflection at a temperature of    800.0000
The in-plane velocity of the surface is  0.0000000E+00
The angular velocity of the surface is  0.0000000E+00
For all molecular species
The gas-surface interaction is diffuse
Rotational energy accomm. coeff.   1.000000
The fraction of specular reflection is  0.0000000E+00
The fraction adsorbed is  0.0000000E+00
Surface          2  is defined by         61  points
Surface          2  is comprised of          1  segments
Segment          1  is a circular arc
The arc is in an anticlockwise direction
 The arc is a circle with center at   1.300000       0.7000000
The segment starts at   1.500000       0.7000000
The segment ends at   1.500000       0.7000000
 The number of sampling property intervals along this segment is       60
 The data on the        60  intervals is in        1  groups
Group          1  is a solid surface containing        60  intervals
The gas-surface interaction is species independent
Diffuse reflection at a temperature of    500.0000
The in-plane velocity of the surface is  0.0000000E+00
The angular velocity of the surface is  0.0000000E+00
For all molecular species
The gas-surface interaction is diffuse
Rotational energy accomm. coeff.   1.000000
The fraction of specular reflection is  0.0000000E+00
The fraction adsorbed is  0.0000000E+00
Surface          3  is defined by          2  points
Surface          3  is comprised of          1  segments
Segment          1  is a straight line
The segment starts at  0.0000000E+00  0.2000000
The segment ends at  0.9000000      1.100000
 The number of sampling property intervals along this segment is        1
 The data on the         1  intervals is in        1  groups
Group          1  is a stream boundary containing        1  intervals
The velocity component in the x direction is   2000.000
The velocity component in the Y direction is  0.0000000E+00
The velocity component in the Z direction is  0.0000000E+00
The gas temperature is    300.0000
```

```
Any vibrational temperature is    300. 0000
The number density is  2.0000000E+20
The fraction of species           1 is  0.2000000
The fraction of species           2 is  0.8000000
The side at the minimum x coord is a stream boundary
The side at the maximum x coord is a stream boundary
The side at the minimum y coord is the axis of symmetry
The side at the maximum y coord is a stream boundary
The stream is the initial state of the flow
No molecules enter from a DSMIF.DAT file
The initial gas is homogeneous at the stream conditions
The sampling is for an unsteady flow
The computational variables may be altered
The reference value of the weighting factor is   100.0000
The number of divisions in x & y directions is changed by a factor of 1.00
The number of elements per division in the x & y directions is         10
The initial number of molecules is changed by a factor of   1.000000
The number of moves in a mean collision time is          5
The number of moves in a mean transit time is          2
The standard sampling interval is changed by a factor of   1.000000
The standard output interval is changed by a factor of   2.000000
A time average is taken over the last  0.500000  of the sampling interval
Employ nearest neighbor collisions for all flows
```

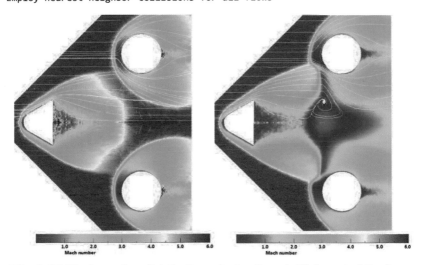

Fig. 8.37 Mach number distributions during the establishment of the flow.

The flow is impulsively started at zero time and some time elapses before there is any interaction between the bow shocks produced by the blunt cone and torus. The first Mach number distribution in Fig. 8.37 shows the flowfield shortly before the commencement of the interaction. The second shows that the interaction causes the flow to choke in the torus and a shock wave moves upstream.

The flow does not come to a steady state at large times. Instead, there is an oscillatory flow in which the ratio of the maximum number to the minimum number of simulated molecules is 5:4. The flow properties at the maximum and minimum number are shown in Figs. 8.38 and 8.39, respectively. The main points of difference are in the location of the forward stagnation point on the torus and the strength of the vortex behind the blunt cone. This vortex has just appeared in Fig. 8.37 and its pulsations eventually drive the oscillatory flow.

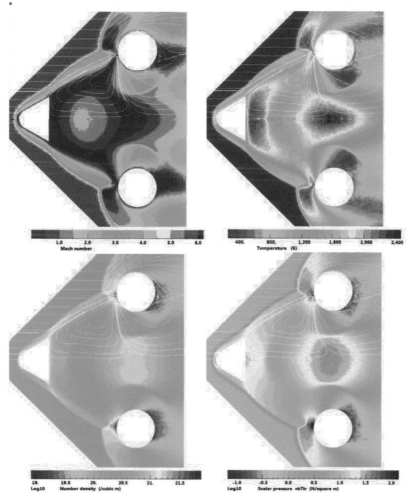

Fig. 8.38 The flow properties at the maximum number of molecules.

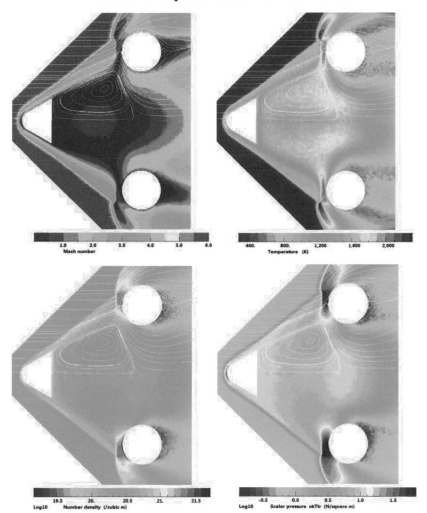

Fig. 8.39 The flow properties at the minimum number of molecules.

The stagnation point on the torus moves through only about 15°, but there is a drastic change in the geometry of the interaction between the two bow shock waves. That at the maximum molecule number is an Edney type IV interaction that involves a jet of supersonic air close to the surface of the toroidal ballute. This leads to the locally high values of the pressure on and the heat transfer to the surface that are shown in Figs. 8.40 and 8.41.

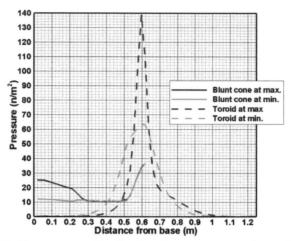

Fig. 8.40 The pressure distributions from rear stagnation points.

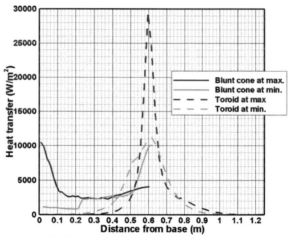

Fig. 8.41 The heat transfer from the rear stagnation points.

Some velocities opposite to the stream in the vortex of the maximum molecule case are supersonic and the consquent pressure on the base is so large that there is an overall thrust of 0.83 N on the blunt cone. This compares with a drag of 1.00 N at the minimum molecule number. The maximum pressures on and heat transfers to the torus are higher than those on the blunt cone, but the minimum pressures and heat transfers are lower. The pressure over the spherical portion of the blunt cone is unaffected by the oscillation, but there is a surprisingly large variation in the heat transfer.

8.8 The effect of sweep on transonic cylinder drag

While there cannot be flow gradients in the z direction of a two-dimensional flow, there can be a uniform velocity in that direction and its magnitude may vary over the x-y plane. The effect of sweep may be studied either through the assignment of a z component to the stream or a velocity in the z direction to the surface. The first option is chosen in this example, but the unswept case is the first to be calculated. This is the flow of air past a circular cylinder at a Mach number of 1.0 and a Knudsen number based on the diameter of the cylinder of approximately 0.007. The data report is:

```
Data summary for program DS2V
 The n in version number n.m is         4
 The m in version number n.m is         5
 The approximate number of megabytes for the calculation is          800
 The flow is two-dimensional
 x limits of flowfield are -0.5000000    ,   0.8000000
 y limits of flowfield are -0.5000000    ,   0.5000000
 Approximate fraction of the bounding rectangle occupied by flow is 1.00000
 Estimated ratio of the average number density to the ref. value is 1.0000
 The number of molecular species is         2
 Maximum number of vibrational modes of any species is          0
 The number of chemical reactions is          0
 The number of surface reactions is          0
 The reference diameter of species          1  is  4.0699999E-10
 The reference temperature of species          1  is    273.0000
 The viscosity-temperature power law of species          1  is  0.7700000
 The reciprocal of the VSS scattering parameter of species  1  is   1.000000
 The molecular mass of species           1  is  5.3120001E-26
 Species           1  is described as Ideal_oxygen
 Species           1  has         2  rotational degrees of freedom
 and the constant relaxation collision number is    5.000000
 The reference diameter of species          2  is  4.1699999E-10
 The reference temperature of species          2  is    273.0000
 The viscosity-temperature power law of species          2  is  0.7400000
 The reciprocal of the VSS scattering parameter of species 2  is    1.000000
 The molecular mass of species           2  is  4.6500001E-26
 Species           2  is described as Ideal_nitrogen
 Species           2  has         2  rotational degrees of freedom
 and the constant relaxation collision number is    5.000000
 The number density of the stream or reference gas is  1.0000000E+21
 The stream temperature is    300.0000
 The velocity component in the x direction is    246.1600
 The velocity component in the y direction is  0.0000000E+00
 The velocity component in the z direction is    246.1600
```

```
The fraction of species       1  is  0.2000000
The fraction of species       2  is  0.8000000
There are         1  separate surfaces
The number of points on surface        1  is         121
The maximum number of points on any surface is       121
The total number of solid surface groups is        1
The total number of solid surface intervals is      120
The total number of flow entry elements is        0
Surface        1  is defined by       121  points
Surface        1  is comprised of       1  segments
Segment        1  is a circular arc
The arc is in an anticlockwise direction
  The arc is a circle with center at  0.0000000E+00  0.0000000E+00
The segment starts at  0.1000000     0.0000000E+00
The segment ends at  0.1000000      0.0000000E+00
  The number of sampling property intervals along this segment is       120
  The data on the        120  intervals is in        1  groups
Group        1  is a solid surface containing      120  intervals
The gas-surface interaction is species independent
Diffuse reflection at a temperature of   300.0000
The in-plane velocity of the surface is  0.0000000E+00
The normal-to-plane velocity of the surface is  0.0000000E+00
For all molecular species, the gas-surface interaction is diffuse
Rotational energy accomm. coeff.   1.000000
The fraction of specular reflection is  0.0000000E+00
The fraction adsorbed is  0.0000000E+00
The side at the minimum x coord is a stream boundary
The side at the maximum x coord is a stream boundary
The side at the minimum y coord is a stream boundary
The side at the maximum y coord is a stream boundary
The stream is the initial state of the flow
No molecules enter from a DSMIF.DAT file
The initial gas is homogeneous at the stream conditions
The sampling is for an eventual steady flow
The calculation employs the standard computational parameters
```

A relatively small flowfield has been chosen for the test calculation and it should be repeated with progressively larger flowfields in order to determine the effect of flowfield size on the flow. The Mach number ditribution at time 0.0029 s after the impulsive insertion of the cylinder is shown in Fig. 8.42. There is a sharp transition from the undisturbed sonic flow to a flow at a Mach number between 0.7 and 0.8. The flow is at approximately five cylinder transit times of the sonic flow. The shock wave has moved about 1.7 cylinder diameters against the sonic flow and therefore has moved at an average speed of about 1.35 times the speed of sound.

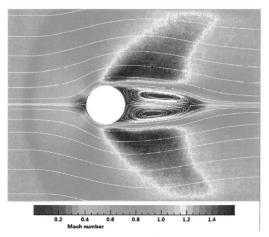

Fig. 8.42 The upstream propagation of the flow disturbance in a sonic stream.

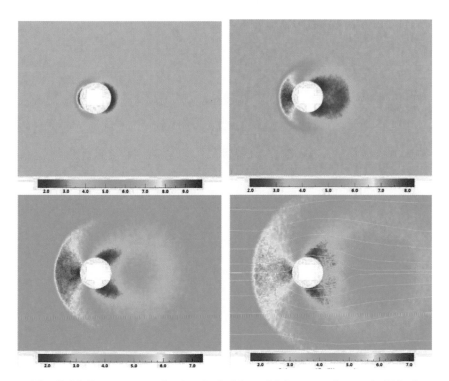

Fig. 8.43 Pressures at the 1st, 2nd, 5th and 9th output intervals (N/m²).

A representative set of pressure distributions over the developing flowfield are shown in Fig. 8.43. The Mach number of the shock wave produced by the instantaneous blocking of a one-dimensional sonic flow in air is 1.766 and the pessure behind this wave would be 14.4 N/m². However, the one-dimensional value is not attained and the pressure at the stagnation point at the time of the first output was 10 N/m². There is then a progressive drop in the pressure and the value at the end of the 9th output interval was just uner 7 N/m².

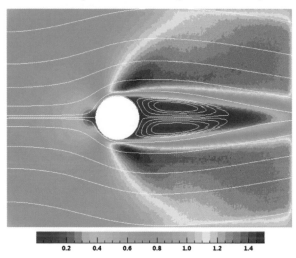

Fig. 8.44 The steady Mach number contours around the unswept cylinder.

The vortices in the separated flow are symmetrical in the time-averaged steady flow that is shown in Fig 8.44. They were slightly asymmetric when the flow was sampled over a shorter time interval. There did not appear to be vortex shedding, but this could change if the computational flowfield was much larger. The force on the body in the stream direction was 1.06 N and this corresponds to a drag coefficient of 1.83.

The unswept calculation was made with the velocity component in the x direction equal to the speed of sound 348.12 m/s. The only change to the data in order to simulate a cylinder swept at 45° is to, as shown in the data description, set this velocity component to 246.16 m/s and the velocity component in the z direction to the same value. The steady flow Mach number contours in this flow are shown in Fig. 8.45.

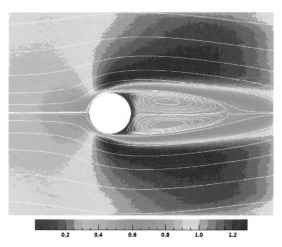

Fig. 8.45 The steady Mach number contours around the 45° swept cylinder.

The force on the swept cylinder in the direction normal to the cylinder is 0.526 N so that the component in the drag drection is 0.372 N.   The shear force along the cylinder is 0.088 N and its component in the drag direction is 0.062 N.   The total drag is therefore 0.434 N and the drag coefficient is 0.75.   The sweep therefore reduces the drag coefficient by 59%.

Fig. 8.46 illustrates how the w velocity components that must not vary in the z direction may vary over the x-y plane.

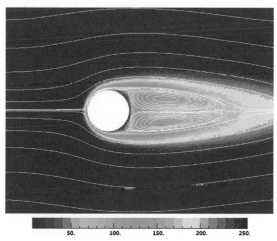

Fig. 8.46 The contours in the x-y plane of the w velocity components (m/s).

8.9 Vacuum pump driven by thermal creep

One of the options for the types of the minimum and maximum x boundaries is "periodic boundaries". Its selection causes the molecules that exit across one of these boundaries to enter across the other boundary. It is employed in this example to study a single stage of a vacuum pump that is operating in the zero pressure difference or maximum flux mode. The pump was invented by Sone (2003) and demonstrates that the thermal creep that occurs only when the mean free paths are comparable with the scale lengths of the temperature gradients can be put to practical use. The data report is:

```
Data summary for program DS2V
  The n in version number n.m is        4
  The m in version number n.m is        5
  The approximate number of megabytes for the calculation is        500
  The flow is two-dimensional
  x limits of flowfield are  0.0000000E+00 ,  0.7500000
  y limits of flowfield are  0.0000000E+00 ,  0.3510000
  The approx. fraction of the bounding rectangle occupied by flow is  0.6670
  The estimated ratio of the average number density to ref. value is    1.00
  The number of molecular species is        1
  Maximum number of vibrational modes of any species is        0
  The number of chemical reactions is        0
  The number of surface reactions is        0
  The reference diameter of species      1  is  4.1699999E-10
  The reference temperature of species      1  is   273.0000
  The viscosity-temperature power law of species      1  is  0.8100000
  The reciprocal of the VSS scattering parameter of species 1  is   1.000
  The molecular mass of species        1  is  6.6300000E-26
  Species        1  is described as Argon
  Species        1  has        0  rotational degrees of freedom
  The number density of the stream or reference gas is  2.0000000E+19
  The stream temperature is    400.0000
  The velocity component in the x direction is  0.0000000E+00
  The velocity component in the y direction is  0.0000000E+00
  The velocity component in the z direction is  0.0000000E+00
  The fraction of species        1  is   1.000000
  There are        1  separate surfaces
  The number of points on surface        1  is        41
  The maximum number of points on any surface is        41
  The total number of solid surface groups is        32
  The total number of solid surface intervals is        40
  The total number of flow entry elements is        0
  Surface        1  is defined by       41  points
  Surface        1  is comprised of        5  segments
  Segment        1  is a straight line
  The segment starts at  0.0000000E+00  0.1750000
  The segment ends at  0.2875000      0.1750000
  The number of sampling property intervals along this segment is        10
  Segment        2  is a straight line
```

```
The segment ends at  0.2875000      0.3500000
The number of sampling property intervals along this segment is          5
Segment        3  is a straight line
The segment ends at  0.4625000      0.3500000
The number of sampling property intervals along this segment is         10
The number of sampling property intervals along this segment is          5
Segment        4  is a straight line
The segment ends at  0.4625000      0.1750000
The number of sampling property intervals along this segment is          5
Segment        5  is a straight line
The segment ends at  0.7500000      0.1750000
The number of sampling property intervals along this segment is         10
The data on the          40  intervals is in          32  groups
Group          1  is a solid surface containing          1  intervals
The gas-surface interaction is species independent
Diffuse reflection at a temperature of    410.0000
The in-plane velocity of the surface is   0.0000000E+00
The normal-to-plane velocity of the surface is   0.0000000E+00
For all molecular species
The gas-surface interaction is diffuse
Rotational energy accomm. coeff.   1.000000
The fraction of specular reflection is   0.0000000E+00
The fraction adsorbed is  0.0000000E+00
Group          2  is a solid surface containing          1  intervals
The gas-surface interaction is species independent
Diffuse reflection at a temperature of    430.0000
The remaining information is the same as for the preceding group.
Group          3  is a solid surface containing          1  intervals
The gas-surface interaction is species independent
Diffuse reflection at a temperature of    450.0000
The remaining information is the same as for the preceding group.
Group          4  is a solid surface containing          1  intervals
The gas-surface interaction is species independent
Diffuse reflection at a temperature of    470.0000
The remaining information is the same as for the preceding group.
Group          5  is a solid surface containing          1  intervals
The gas-surface interaction is species independent
Diffuse reflection at a temperature of    490.0000
The fraction adsorbed is  0.0000000E+00
The remaining information is the same as for the preceding group.
Group          6  is a solid surface containing          1  intervals
The gas-surface interaction is species independent
Diffuse reflection at a temperature of    510.0000
The remaining information is the same as for the preceding group.
The remaining information is the same as for the preceding group.
Group          7  is a solid surface containing          1  intervals
The gas-surface interaction is species independent
Diffuse reflection at a temperature of    530.0000
The remaining information is the same as for the preceding group.
Group          8  is a solid surface containing          1  intervals
The gas-surface interaction is species independent
Diffuse reflection at a temperature of    550.0000
The remaining information is the same as for the preceding group.
Group          9  is a solid surface containing          1  intervals
The gas-surface interaction is species independent
```

```
Diffuse reflection at a temperature of   570.0000
The remaining information is the same as for the preceding group.
Group          10  is a solid surface containing        1  intervals
The gas-surface interaction is species independent
Diffuse reflection at a temperature of   590.0000
The remaining information is the same as for the preceding group.
Group          11  is a solid surface containing        5  intervals
The gas-surface interaction is species independent
Diffuse reflection at a temperature of   600.0000
The remaining information is the same as for the preceding group.
Group          12  is a solid surface containing        1  intervals
The gas-surface interaction is species independent
Diffuse reflection at a temperature of   580.0000
The remaining information is the same as for the preceding group.
Group          13  is a solid surface containing        1  intervals
The gas-surface interaction is species independent
Diffuse reflection at a temperature of   540.0000
The remaining information is the same as for the preceding group.
Group          14  is a solid surface containing        1  intervals
The gas-surface interaction is species independent
Diffuse reflection at a temperature of   500.0000
The remaining information is the same as for the preceding group.
Group          15  is a solid surface containing        1  intervals
The gas-surface interaction is species independent
Diffuse reflection at a temperature of   460.0000
The remaining information is the same as for the preceding group.
Group          16  is a solid surface containing        1  intervals
The gas-surface interaction is species independent
Diffuse reflection at a temperature of   420.0000
The remaining information is the same as for the preceding group.
Group          17  is a solid surface containing        1  intervals
The gas-surface interaction is species independent
Diffuse reflection at a temperature of   380.0000
The remaining information is the same as for the preceding group.
Group          18  is a solid surface containing        1  intervals
The gas-surface interaction is species independent
Diffuse reflection at a temperature of   340.0000
The remaining information is the same as for the preceding group.
Group          19  is a solid surface containing        1  intervals
The gas-surface interaction is species independent
Diffuse reflection at a temperature of   300.0000
The remaining information is the same as for the preceding group.
Group          20  is a solid surface containing        1  intervals
The gas-surface interaction is species independent
Diffuse reflection at a temperature of   260.0000
The remaining information is the same as for the preceding group.
Group          21  is a solid surface containing        1  intervals
The gas-surface interaction is species independent
Diffuse reflection at a temperature of   220.0000
The remaining information is the same as for the preceding group.
Group          22  is a solid surface containing        5  intervals
The gas-surface interaction is species independent
Diffuse reflection at a temperature of   200.0000
The remaining information is the same as for the preceding group.
Group          23  is a solid surface containing        1  intervals
```

```
The gas-surface interaction is species independent
Diffuse reflection at a temperature of    210.0000
The remaining information is the same as for the preceding group.
Group          24  is a solid surface containing          1  intervals
The gas-surface interaction is species independent
Diffuse reflection at a temperature of    230.0000
The remaining information is the same as for the preceding group.
Group          25  is a solid surface containing          1  intervals
The gas-surface interaction is species independent
Diffuse reflection at a temperature of    250.0000
The remaining information is the same as for the preceding group.
Group          26  is a solid surface containing          1  intervals
The gas-surface interaction is species independent
Diffuse reflection at a temperature of    270.0000
The remaining information is the same as for the preceding group.
Group          27  is a solid surface containing          1  intervals
The gas-surface interaction is species independent
Diffuse reflection at a temperature of    290.0000
The remaining information is the same as for the preceding group.
Group          28  is a solid surface containing          1  intervals
The gas-surface interaction is species independent
Diffuse reflection at a temperature of    310.0000
The remaining information is the same as for the preceding group.
Group          29  is a solid surface containing          1  intervals
The gas-surface interaction is species independent
Diffuse reflection at a temperature of    330.0000
The remaining information is the same as for the preceding group.
Group          30  is a solid surface containing          1  intervals
The gas-surface interaction is species independent
Diffuse reflection at a temperature of    350.0000
The remaining information is the same as for the preceding group.
Group          31  is a solid surface containing          1  intervals
The gas-surface interaction is species independent
Diffuse reflection at a temperature of    370.0000
The remaining information is the same as for the preceding group.
Group          32  is a solid surface containing          1  intervals
The gas-surface interaction is species independent
Diffuse reflection at a temperature of    390.0000
The remaining information is the same as for the preceding group.
The side at the minimum x coord is a periodic boundary
The side at the maximum x coord is a periodic boundary
The side at the minimum y coord is a plane of symmetry
The side at the maximum y coord is not in the flow
The stream is the initial state of the flow
No molecules enter from a DSMIF.DAT file
The initial gas is homogeneous at the stream conditions
The sampling is for an eventual steady flow
 Number of divisions in x & y directions is changed by factor of  0.4000
The number of elements per division in the x & y directions is          10
The initial number of molecules is changed by a factor of   1.000000
The number of moves in a mean collision time is        5
The number of moves in a mean transit time is        2
The standard sampling interval is changed by a factor of   1.000000
The standard output interval is changed by a factor of   1.000000
Employ nearest neighbor collisions for all flows
```

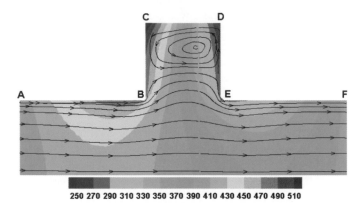

250 270 290 310 330 350 370 390 410 430 450 470 490 510

Fig. 8.47 The temperature (K) contours.

As noted earlier, the ends of the bounding rectangle are periodic boundaries. The lower boundary is a plane of symmetry, while **ABCDEF** is a diffusely reflecting surface within the bounding rectangle. There is a linear temperature gradient between 400 K at **A** and 600 K at **B**. The segment **BC** is at a uniform 600 K and there is a linear drop to 200 K at **D**. The segment **DE** is at a uniform 200 K and the temperature then increases linearly to 400 K at **F** to match the temperature at **A**.

The temperature contours in the gas are shown in Fig. 8.47 and range from just over 240 K to just under 530 K. The large temperature slips are indicative of the Knudsen number which is 0.42 based on the semi-height of the channel or 0.21 based on the full height. These are typical of the Knudsen numbers at which thermal creep is strongest. The variation in scalar pressure is from 0.095 N/m^2 to 0.115 N/m^2 and, in comparison with the temperature variation, relatively small. The effect of thermal creep is to produce a flow velocity from cold regions to hot regions. It therefore promotes a flow in the positive x direction between **A** and **B** and a flow in the negative x direction between **E** and **F**. The negative temperature gradient between **D** and **C** also promotes a flow in the negative direction but, because a vortex forms in the side lobe, it actually promotes a flow in the positive x direction. The flow speed contours are shown in Fig. 8.48. The maximum flow speed is approximately 11 m/s and the overall mas flux is equivalent to a speed of about 8 m/s at the average flow density. Should the flow be blocked, a pressure difference develops and there can be a cascade of many similar stages.

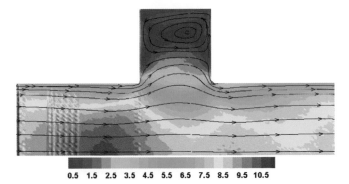

Fig. 8.48 The flow speed (m/s) contours.

The graphics in the DS2V program has difficulties in drawing streamlines at the very low speeds in the side lobe and a Tecplot output file option that is generally deactivated was employed for the generation of these figures.   The generation of streamlines is facilitated by the more regular cells before cell adaption, but the source of numerical artifacts near the flow inlet is unclear. Given the overall speed of about 8 m/s and the very high temperature gradient along the surface segment **CD** the magnitude of the flow speeds in the side lobe is surprisingly small. The flow speed contours in this region of the flow are shown in Fig. 8.49.

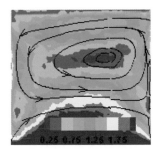

Fig. 8.49 The flow speed (m/s) contours in the side lobe.

The flow speeds in the vortex do not exceed 1.75 m/s and the inner streamline is well defined even though the flow speed is close to 0.25 m/s. The flow quantities are based on sample sizes of approximately 5×10^7 and this required a run of several days on a 3 GHz Intel I7 CPU. **As already noted in §8.4, the frequent claims that DSMC cannot be applied to very low speed flows are now outdated.**

8.10 Free jet in cryogenically pumped chamber

This is a relatively simple example that illustrates how DSMC can be used to simulate laboratory and industrial vacuum systems. An axially symmetric vacuum chamber with typical dimensions of one metre is pumped by a cryogenic panel and includes a one cm radius sonic free jet of nitrogen. The data is reported by DS2VD.TXT as:

```
Data summary for program DS2V
 The n in version number n.m is          4
 The m in version number n.m is          5
 The approximate number of megabytes for the calculation is        200
 The flow is axially symmetric
 There are no radial weighting factors
 x limits of flowfield are -1.0000000E-03 ,  1.101000
 y limits of flowfield are  0.0000000E+00 ,  0.8010000
 The approximate fraction of the bounding rectangle occupied by flow is 0.8
 Estimated ratio of the average number density to reference value is 1.0
 The number of molecular species is        1
 Maximum number of vibrational modes of any species is          0
 The number of chemical reactions is        0
 The number of surface reactions is        0
 The reference diameter of species        1  is  4.1699999E-10
 The reference temperature of species        1  is   273.0000
 The viscosity-temperature power law of species        1  is  0.7400000
 The reciprocal of the VSS scattering parameter of species 1  is   1.000000
 The molecular mass of species          1  is  4.6500001E-26
 Species        1  is described as Nitrogen
 Species        1  has        2  rotational degrees of freedom
 and the constant relaxation collision number is   5.000000
 The number density of the stream or reference gas is  2.0000000E+19
 The stream temperature is   300.0000
 The velocity component in the x direction is  0.0000000E+00
 The fraction of species        1  is   1.000000
 There are        1  separate surfaces
 The number of points on surface        1  is         84
 The maximum number of points on any surface is        84
 The total number of solid surface groups is        4
 The total number of solid surface intervals is        82
 The total number of flow entry elements is        1
 Surface        1  is defined by        84  points
 Surface        1  is comprised of        9  segments
 Segment        1  is a straight line
 The segment starts at  0.4000000       0.0000000E+00
 The segment ends at  0.4000000       9.9999998E-03
  The number of sampling property intervals along this segments          1
 Segment          2  is a straight line
```

```
The segment ends at   0.0000000E+00  0.2000000
 The number of sampling property intervals along this segment is         5
Segment           3  is a straight line
The segment ends at   0.0000000E+00  0.4000000
 The number of sampling property intervals along this segment is         5
Segment           4  is a circular arc in a clockwise direction
  The arc is a circle with center at  0.4000000        0.4000000
The segment ends at   0.4000000       0.8000000
 The number of sampling property intervals along this segment is        12
Segment           5  is a straight line
The segment ends at   0.9000000       0.8000000
 The number of sampling property intervals along this segment is        12
Segment           6  is a circular arc in a clockwise direction
  The arc is a circle with center at  0.9000000        0.6000000
The segment ends at    1.100000       0.6000000
 The number of sampling property intervals along this segment is         6
Segment           7  is a straight line
The segment ends at    1.100000       0.2000000
 The number of sampling property intervals along this segment is         6
Segment           8  is a straight line
The segment ends at   0.7000000       0.2000000
 The number of sampling property intervals along this segment is         6
Segment           9  is a straight line
The segment ends at   0.7000000       0.0000000E+00
 The number of sampling property intervals along this segment is        30
  The data on the           83  intervals is in          4  groups
Group            1  is an entry line containing          1  intervals
The velocity component in the x direction is    353.0000
The velocity component in the Y direction is   0.0000000E+00
The velocity component in the Z direction is   0.0000000E+00
The gas temperature is    300.0000
Any vibrational temperature is    300.0000
The number density is   9.9999998E+22
The fraction of species           1  is    1.000000
The flow commences at time   0.0000000E+00
The flow ceases at time   1.0000000E+20
The interval is a plane of symmetry when the flow is stopped
Group            2  is a solid surface containing         22  intervals
The gas-surface interaction is species independent
Diffuse reflection at a temperature of    300.0000
The in-plane velocity of the surface is   0.0000000E+00
The angular velocity of the surface is   0.0000000E+00
The gas-surface interaction is diffuse
Rotational energy accomm. coeff.   1.000000
The fraction of specular reflection is   0.0000000E+00
The fraction adsorbed is   0.0000000E+00
Group            3  is a solid surface containing         12  intervals
The gas-surface interaction is species independent
```

```
Diffuse reflection at a temperature of    30.00000
The in-plane velocity of the surface is  0.0000000E+00
The angular velocity of the surface is  0.0000000E+00
The gas-surface interaction is diffuse
Rotational energy accomm. coeff.   1.000000
The fraction of specular reflection is  0.0000000E+00
The fraction adsorbed is  0.9800000
Group           4  is a solid surface containing          48  intervals
The gas-surface interaction is species independent
Diffuse reflection at a temperature of    300.0000
The in-plane velocity of the surface is  0.0000000E+00
The angular velocity of the surface is  0.0000000E+00
The gas-surface interaction is diffuse
Rotational energy accomm. coeff.   1.000000
The fraction of specular reflection is  0.0000000E+00
The fraction adsorbed is  0.0000000E+00
The side at the minimum x coord is not in the flow
The side at the maximum x coord is not in the flow
The side at the minimum y coord is the axis of symmetry
The side at the maximum y coord is not in the flow
The stream is the initial state of the flow
No molecules enter from a DSMIF.DAT file
The initial gas is homogeneous at the stream conditions
The sampling is for an eventual steady flow
The calculation employs the standard computational parameters
```

The vacuum chamber is defined by a single surface with nine segments in four groups and 83 sampling intervals. It is an open surface that begins and ends on the lower boundary at the axis. Group one is a single segment defining the disc at the centre-line of the chamber that forms the entry plane of the free jet. The second group is the three solid surface segments to the left of the free jet entry. The third group defines the cylindrical cryogenic panel along the. outer edge of the chamber. The four solid surface segments on the right of the chamber comprise the final group.

The ratio of the free jet pressure to the initial background pressure is such that a high Mach number is attained in the expansion. The number density contours in Fig. 8.50 show that the pumping rate of the cryogenic panel, that is assumed to be at 30 K with 98% adsorption, is sufficiently high for the high Mach number to be sustained when a steady state is attained. The Mach number contours in Fig. 8.43 show an expansion to Mach 7.5 before a normal shock wave forms ahead of the target surface. The pressure on this surface ranges from 2.6 N/m^2 at the axis to 0.77 N/m^2.

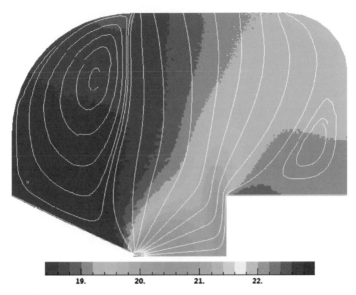

Fig. 8.50 Number density contours in the steady state.

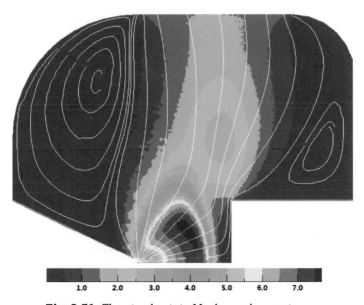

Fig. 8.51 The steady state Mach number contours.

8.11 Rotating planet

The DSMC method can be applied to planetary and astrophysical problems with very large physical dimensions as long as the number density is sufficiently low for the Knudsen number to be in the range that is currently feasible. This calculation is for a sphere with a diameter of 10,000 m rotating with an angular speed of 0.3 rad/s. Both the undisturbed gas temperature and the temperature of the diffusely reflecting surface are at 300 K. The undisturbed number density of the hard sphere gas is 10^{16} m^{-3} so that the mean free path is 141 m and the Knudsen number based on the diameter is 0.0141. The peripheral speed of the surface at the equator is 1500 m/s and this corresponds to a speed ratio of just over four.

The **DS2V** program is a 32 bit program with only about six digit accuracy. This would lead to insuperable problems for applications with very small linear dimensions and the program includes a scale factor that increases the characteristic linear dimension to the order of 1 to 10 m. This permits the introduction of a variable **DCRIT** that is larger than any round-off error and can be used in tests for floating point equality. However, linear dimensions that are significantly larger than 10 m are not scaled down and, for this application, the value of **DCRIT** had to be increased within the source code.

The data file was reported in **DS2VD.TXT** as:

```
Data summary for program DS2V
 The n in version number n.m is          4
 The m in version number n.m is          5
 The approximate number of megabytes for the calculation is        400
 The flow is axially symmetric. There are radial weighting factors
 x limits of flowfield are  -20000.00     ,    20000.00
 y limits of flowfield are  0.0000000E+00 ,    30000.00
 The approx. fraction of the bounding rectangle occupied by flow is  0.9500
 The estimated ratio of the average number density to ref. value is 1.00
 The number of molecular species is          1
 Maximum number of vibrational modes of any species is          0
 The number of chemical reactions is          0
 The number of surface reactions is          0
 The reference diameter of species          1 is  4.0000001E-10
 The reference temperature of species          1 is    273.0000
 The viscosity-temperature power law of species          1 is  0.5000000
 The reciprocal of the VSS scattering parameter of species  1 is    1.00
 The molecular mass of species          1 is  5.0000001E-26
 Species          1 is described as Hard_sphere_gas
```

```
Species             1  has             0  rotational degrees of freedom
The number density of the stream or reference gas is  1.0000000E+16
The stream temperature is    300.0000
The velocity component in the x direction is  0.0000000E+00
The fraction of species         1  is    1.000000
There are           1  separate surfaces
The number of points on surface         1  is            91
The maximum number of points on any surface is            91
The total number of solid surface groups is            1
The total number of solid surface intervals is            90
The total number of flow entry elements is            0
Surface             1  is defined by       91  points
Surface             1  is comprised of        1  segments
Segment             1  is a circular arc in an anticlockwise direction
 The arc is a circle with center at  0.0000000E+00  0.0000000E+00
The segment starts at     5000.000        0.0000000E+00
The segment ends at   -5000.000        0.0000000E+00
 The number of sampling property intervals along this segment is        90
 The data on the            90  intervals is in        1  groups
Group             1  is a solid surface containing        90  intervals
The gas-surface interaction is species independent
Diffuse reflection at a temperature of 300.0000
The in-plane velocity of the surface is  0.0000000E+00
The angular velocity of the surface is  0.3000000
For all molecular species  The gas-surface interaction is diffuse
Rotational energy accomm. coeff.    1.000000
The fraction of specular reflection is  0.0000000E+00
The fraction adsorbed is  0.0000000E+00
The side at the minimum x coord is a stream boundary
The side at the maximum x coord is a stream boundary
The side at the minimum y coord is the axis of symmetry
The side at the maximum y coord is a stream boundary
The stream is the initial state of the flow
The initial gas is homogeneous at the stream conditions
The sampling is for an eventual steady flow
The calculation employs the standard computational parameters
```

Figure 8.52 shows that the maximum flow speed is just over 1,000 m/s and is therefore much less than the peripheral speed of the surface at the equator.   Multiple runs with progressively larger flowfields are required in order to determine when the flowfield is sufficiently large to not have any effect on the flow near the cylinder. For this calculation, the flow speed at the boundaries is about 10 m/s at the axis, 2-3 m/s along the sides, 1 m/s at the corners and 5 m/s at the intersection of the equatorial plane and the outer boundaries. These values are small in comparison with the 40 m/s at the centre of the vortices and 100 m/s in the equatorial plane at the radius of vortex centres. The flowfield is probably sufficiently large, but this should be verified by calculations with even larger flowfields.

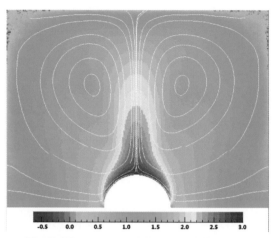

Fig. 8.52 Contours of flow speed ($\log_{10}$ m/s).

The maximum speed is three times the speed of sound at 300 K, but the temperature increases to more than 900K and the maximum Mach number in the flow is 1.63. While it appears in Fig. 8.53 that there is a smooth transition from supersonic to subsonic flow, the pressure contours in Fig. 8.54 show a weak annular shock wave.

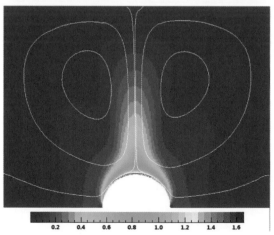

Fig. 8.53 Mach number contours.

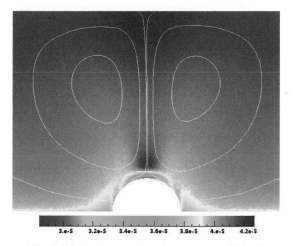

Fig. 8.54 Contours of scalar pressure (N/m^2).

Over most of the flowfield, the circumferential component makes the dominant contribution to the speed. Contours of the radial component are plotted in Fig. 8.47 and these show that the maximum radial component occurs near the equatorial plane beyond the weak shock wave. However, even in this region, the circumferential component is about 20 % higher than the radial component.

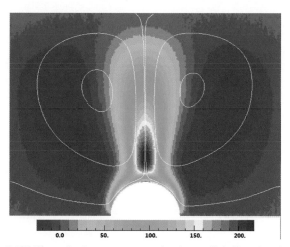

Fig. 8.55 The velocity component in the radial direction (m/s).

Most astrophysical applications require a gravitational field. This is a straightforward extension that was treated in Bird (1994).

8.12 Spherical Taylor-Couette flow

Cylindrical Taylor-Couette flow involves a rotating inner cylinder and a stationary outer cylinder. It was a new class of test case for DSMC in Bird (1994) in that it is an unstable flow and, while the steady state vortex patterns are repeatable at high Knudsen numbers, the unsteady phase is not repeatable. At sufficiently low Knudsen numbers, the flow becomes permanently chaotic. This example involves rotating spheres rather than cylinders and, like the §8.11 example, it has planetary implications.

The radius of the outer sphere is 20% higher than that of the inner sphere and the circumferential speed of this sphere at the equator is 244 m/s. This speed is two thirds the speed of sound in the undisturbed hard sphere gas at 300 K. The density is such that the Knudsen based on the gap between the spheres is 0.005.

```
Data summary for program DS2V
 The n in version number n.m is          4
 The m in version number n.m is          5
 The approximate number of megabytes for the calculation is        30
 The flow is axially symmetric
 There are no radial weighting factors
 x limits of flowfield are  -1.250000    ,   1.250000
 y limits of flowfield are  0.0000000E+00 ,   1.250000
 The approx. fraction of the bounding rectangle occupied by flow is 0.1500
 Estimated ratio of the average number density to the ref. value is 1.00000
 The number of molecular species is          1
 Maximum number of vibrational modes of any species is          0
 The number of chemical reactions is          0
 The number of surface reactions is          0
 The reference diameter of species          1 is  4.0000001E-10
 The reference temperature of species          1 is   273.0000
 The viscosity-temperature power law of species          1 is  0.5000000
 The reciprocal of the VSS scattering parameter of species 1 is  1.0000
 The molecular mass of species          1 is  5.0000001E-26
 Species          1 is described as Hard_sphere_gas
 Species          1 has          0 rotational degrees of freedom
 The number density of the stream or reference gas is  1.4067000E+21
 The stream temperature is   300.0000
 The velocity component in the x direction is  0.0000000E+00
 The fraction of species          1 is   1.000000
 There are          2 separate surfaces
 The number of points on surface          1 is        101
 The number of points on surface          2 is        101
 The maximum number of points on any surface is        101
```

```
The total number of solid surface groups is          1
The total number of solid surface intervals is        200
The total number of flow entry elements is          0
Surface          1  is defined by          101  points
Surface          1  is comprised of          1  segments
Segment          1  is a circular arc
The arc is in an anticlockwise direction
The arc is a circle with center at   0.0000000E+00  0.0000000E+00
The segment starts at    1.000000      0.0000000E+00
The segment ends at   -1.000000      0.0000000E+00
The number of sampling property intervals along this segment is        100
The data on the          100  intervals is in          1  groups
Group          1  is a solid surface containing        100  intervals
The gas-surface interaction is species independent
Diffuse reflection at a temperature of    300.0000
The in-plane velocity of the surface is   0.0000000E+00
The angular velocity of the surface is    244.0000
For all molecular species The gas-surface interaction is diffuse
Rotational energy accomm. coeff.   1.000000
The fraction of specular reflection is   0.0000000E+00
The fraction adsorbed is   0.0000000E+00
Surface          2  is defined by          101  points
Surface          2  is comprised of          1  segments
Segment          1  is a circular arc
The arc is in a clockwise direction
 The arc is a circle with center at   0.0000000E+00  0.0000000E+00
The segment starts at   -1.200000      0.0000000E+00
The segment ends at    1.200000      0.0000000E+00
 The number of sampling property intervals along this segment is        100
 The data on the          100  intervals is in          1  groups
Group          1  is a solid surface containing        100  intervals
The gas-surface interaction is species independent
Diffuse reflection at a temperature of    300.0000
The in-plane velocity of the surface is   0.0000000E+00
The angular velocity of the surface is   0.0000000E+00
For all molecular species The gas-surface interaction is diffuse
Rotational energy accomm. coeff.   1.000000
The fraction of specular reflection is   0.0000000E+00
The fraction adsorbed is   0.0000000E+00
The side at the minimum x coord is not in the flow
The side at the maximum x coord is not in the flow
The side at the minimum y coord is the axis of symmetry
The side at the maximum y coord is not in the flow
The stream is the initial state of the flow
No molecules enter from a DSMIF.DAT file
The initial gas is homogeneous at the stream conditions
The sampling is for an eventual steady flow
The calculation employs the standard computational parameters
```

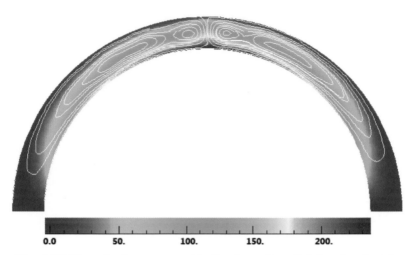

Fig. 8.56 The velocity component in the circumferential direction (m/s).

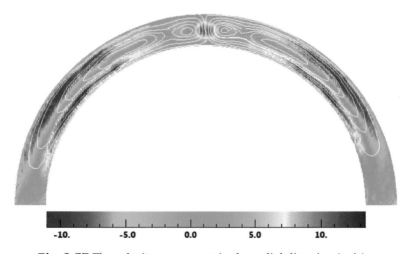

Fig. 8.57 The velocity component in the radial direction (m/s).

The velocity components in the circumferential and radial directions are shown in Figs. 8.56 and 8.57, respectively. The flow is nearly symmetrical about the equator and a single vortex with two cores forms in each hemisphere. The first core near the equator is only slightly elongated while the weaker second core extends through an arc of about 60°. The velocity components in an azimuthal plane are of the order of one tenth the circumferential velocity components.

8.13 Two-dimensional under-expanded jet

DSMC is much more robust than continuum CFD and, in order to study complex flow configurations, it is sometimes applied to continuum flows at densities such that the requirement for the mean collisional separation to be small in comparison with the mean free path cannot be met. The average value of the mcs/mfp ratio in this example is 1.65 and, at this value, the heat transfer to any surfaces would be in error by about 15%. However, there are no surfaces of interest in this application and the overall flow configuration would be in agreement with observations of real flows.

The application is the two-dimensional expansion of a sonic free-jet into a stationary gas at one fifth the entry pressure of the jet. Both the jet and gas are monatomic and the gas temperature is equal to the jet entry temperature. The Knudsen number based on the full height of the jet is 0.00047. The data file was reported in **DS2VD.TXT** as:

Data summary for program DS2V
The n in version number n.m is   4
The m in version number n.m is   5
The approximate number of megabytes for the calculation is   600
The flow is two-dimensional
x limits of flowfield are 0.0000000E+00 ,   4.500000
y limits of flowfield are 0.0000000E+00 ,   1.000000
The approx. fraction of bounding rectangle occupied by flow is 1.00
Estimated ratio of the average number density to the ref. value is 2.0
The number of molecular species is   1
Maximum number of vibrational modes of any species is   0
The number of chemical reactions is   0
The number of surface reactions is   0
The reference diameter of species   1 is 4.0000001E-10
The reference temperature of species   1 is   273.0000
The viscosity-temperature power law of species   1 is 0.5000000
The reciprocal of the VSS scattering parameter of species 1 is 1.000
The molecular mass of species   1 is 5.0000001E-26
Species   1 is described as Hard_sphere_gas
Species   1 has   0 rotational degrees of freedom
The number density of the stream or reference gas is 2.0000E+21
The stream temperature is   300.0000
The velocity component in the x direction is 0.0000000E+00

The velocity component in the y direction is 0.0000000E+00
The velocity component in the z direction is 0.0000000E+00
The fraction of species   1 is   1.000000
There are   1 separate surfaces
The number of points on surface   1 is   3
The maximum number of points on any surface is   3
The total number of solid surface groups is   2
The total number of solid surface intervals is   1
The total number of flow entry elements is   1
Surface   1 is defined by   3 points
Surface   1 is comprised of   2 segments
Segment   1 is a straight line
The segment starts at 0.1500000   0.0000000E+00
The segment ends at 0.1500000   0.2000000
 The number of sampling property intervals along this segment is   1
Segment   2 is a straight line
The segment ends at 0.0000000E   +00 0.2000000
 The number of sampling property intervals along this segment is   1
 The data on the   2 intervals is in   2 groups
Group   1 is an entry line containing   1 intervals
The velocity component in the x direction is   372.0000
The velocity component in the Y direction is 0.0000000E+00
The velocity component in the Z direction is 0.0000000E+00
The gas temperature is   300.0000
Any vibrational temperature is   300.0000
The number density is 9.9999998E+21
The fraction of species   1 is   1.000000
The flow commences at time 0.0000000E+00
The flow ceases at time 1.0000000E+20
The interval is a plane of symmetry when the flow is stopped
Group   2 is a solid surface containing   1 intervals
The gas-surface interaction is species independent
Diffuse reflection at a temperature of   300.0000
The in-plane velocity of the surface is 0.0000000E+00
The normal-to-plane velocity of the surface is 0.0000000E+00
For all molecular species
The gas-surface interaction is diffuse
Rotational energy accomm. coeff.   1.000000
The fraction of specular reflection is   1.000000

The fraction adsorbed is 0.0000000E+00
The side at the minimum x coord is a stream boundary
The side at the maximum x coord is a stream boundary
The side at the minimum y coord is a plane of symmetry
The side at the maximum y coord is a stream boundary
The stream is the initial state of the flow
No molecules enter from a DSMIF.DAT file
The initial gas is homogeneous at the stream conditions
The sampling is for an eventual steady flow
The calculation employs the standard computational parameters

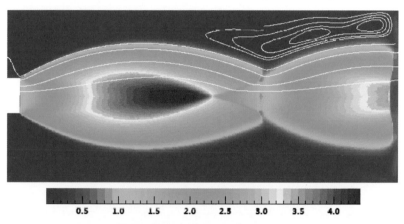

Fig. 8.58 The Mach number contours.

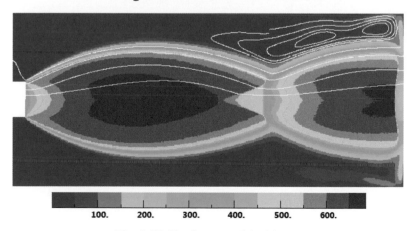

Fig. 8.59 The flow speed (m/s) contours.

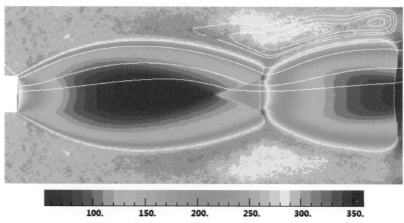

Fig. 8.60 The temperature (K) contours.

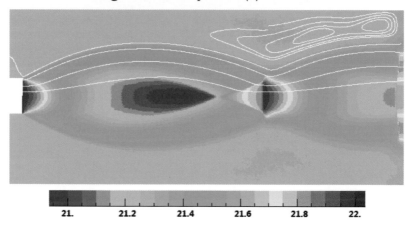

Fig. 8.61 The number density ($\log_{10}$ m^{-3}) contours.

The Mach number, speed, temperature and number density contours are shown in Figs. 8.58 to 8.61. The extent of the flowfield is sufficient to capture one and a half shock diamonds and the Knudsen number is sufficiently small for the shock waves in the Mach reflection to be sharply defined. The streamline from just outside the nozzle lip does not penetrate to the supersonic region before this region is terminated by a normal shock wave just upstream of the flow boundary. The Knudsen number of the flow is sufficiently low for the DSMC calculation to provide a good representation of the continuum flowfield configuration.

The pressure in the stationary background gas is one fifth that at the sonic entry and almost one tenth the stagnation pressure of the jet. There would be some loss of stagnation pressure in the shock waves, but it is surprising that the background stream molecules entering the downstream boundary are sufficient to trigger a normal shock. The supersonic region of the jet would not be affected upstream of the shock, but the subsonic region is deflected outwards and a vortex develops in the background gas.

The background gas is entrained by the jet and there would almost certainly have been less boundary interference if the background gas had been given a velocity component in the x direction of about 100 m/s. At the same time, no boundary condition that is uniform along the boundary can really be satisfactory. The downstream boundary, in particular, requires **adaptive boundary conditions** that vary along its length. Automatically adapting boundaries would be dangerous and, in this case, adapting to the post shock conditions would be counter- productive. However, manual adjustment that took account of the physics could solve the problem and could be efficiently implemented in an interactive program like **DS2V**. One would simply click on a point within the flow upstream of the shock and, with another mouse click, transfer its properties to the boundary. This option will be investigated in the 2 and 3-D extensions of **DSMC.F90**.

References

Bird, G. A. (1994) *Molecular Gas Dynamics and the Direct Simulation of Gas Flows,* Clarendon Press, Oxford.

Lofthouse, A. J., Boyd, I. D. and Wright, M.J. (2007). Effects of continuum breakdown on hypersonic aerothermodynamics. *Phys. Fluids* **19**, 027105.

Sone, Y. and Sugimoto, W. (2003). In *Rarefied Gas Dynamics*, (Ed. A.Ketsdever and E. P. Muntz), p. 1041, AIP, New York.

APPENDIX A. Representative gas properties

The calculations have employed the same gas properties that were listed in Bird (1994). These were largely based on the data for the transport properties that was tabulated in Chapman and Cowling (1970) and, as noted in §4.2, should now be reviewed, updated and extended. At the same time, there is a case for using consistent values over the years and the data that is necessary for the implementation of the VHS and VSS molecular models for some common gases is listed here for quick reference.

Table A.1 The information that is required to implement the VHS model.

Gas	Rotational degrees of freedom	Molecular mass $(10^{-27}\,Kg)$	Diameter at 273 K $(10^{-10}\,m)$	Viscosity-temperature index
Hydrogen H_2	2	3.34	2.92	0.67
Helium He	0	6.65	2.33	0.66
Neon Ne	0	33.5	2.77	0.66
Carbon monoxide CO	2	46.6	4.19	0.73
Nitrogen N_2	2	46.5	4.17	0.74
Nitric oxide NO	2	49.9	4.20	0.79
Oxygen O_2	2	53.1	4.07	0.77
Argon Ar	0	66.3	4.17	0.81
Carbon dioxide CO_2	3	73.1	5.62	0.93
Nitrous oxide N_2O	2	73.1	5.71	0.94
Chlorine Cl_2	2	117.7	6.98	1.01
Krypton Kr	0	139.1	4.76	0.8
Xenon Xe	0	218	5.74	0.85

Table A.2 The additional information that is required for the VSS model.

Gas	VSS deflection parameter α	Diameter at 273 K $(10^{-10}$ m$)$
Hydrogen H_2	1.35	2.88
Helium He	1.26	2.30
Methane CH_4	1.60	4.78
Neon Ne	1.31	2.72
Carbon monoxide CO	1.49	4.12
Nitrogen N_2	1.36	4.11
Oxygen O_2	1.40	4.01
Hydrogen chloride HCl	1.59	5.59
Argon Ar	1.40	4.11
Carbon dioxide CO_2	1.61	5.45
Krypton Kr	1.32	4.70
Xenon Xe	1.44	5.65

The information in Table A.2 is again based on the Chapman and Cowling transport property data, using Eqn. (3.20) for the diameter and Eqn. (3.21) for α .

Table A.3 The characteristic temperatures of diatomic molecules.

Gas	Rotation Θ_{rot} (K)	Vibration Θ_{vib} (K)	Dissociation Θ_{diss} (K)	Ionization Θ_i (K)
CH	20.4	4,102	40,300	129,000
Cl_2	0.35	801	28,700	
CO	2.77	3,103	129,000	162,000
H_2	85.4	6,159	52,000	179,000
N_2	2.88	3,371	113,500	181,000
NO	2.44	2,719	75,500	108,000
O_2	2.07	2,256	59,500	142,000
OH	26.6	5,360	50,900	

The collision numbers for rotational relaxation are generally in the range 1 to 10 and increase with temperature. The actual value has little effect in most flows and a collision number of 5 has been used throughout the examples in this book. There is provision in the programs to employ, instead, a first order polynomial in temperature.

The collision number for vibrational relaxation is assumed to be unity at the characteristic dissociation temperature. Equation (3.34) then permits the collision number at any temperature to be calculated from a single value at a specified temperature.

Table A.4 Typical vibrational relaxation collision numbers at the characteristic vibrational temperature.

Molecule	Collision partner	Collision number
N_2	N_2	52,600
O_2	O_2	17,900
NO	NO	1,400
CO	CO	32.700
CO	Ar	170,000
CO	He	3,900
CO	H_2	900
O_2	Ar	33,100
O_2	He	770
O_2	H_2	230
Cl_2	Cl_2	550

The data in Table A.4 was calculated from the data in Bird (1994) that was based on the experimentally based information in Park (1990). There are very large differences between the values for different gas species and, more disturbingly, similar differences for the same gas with different collision partners. There are unknown errors in the experimental data and, for many gas combinations of interest, no data is available. The gas models that are associated with this book have employed a nominal vibrational relaxation rate that does not depend on the species of the collision partner. The models could be modified to include a collision partner dependent rate, but the lack of data means that most of values would have to be estimated.

Information on the accessible electronic levels is required if electronic energy is to be taken into account and, for temperatures in the range 10,000 K to 20,000 K, the degeneracies and energy levels have been taken from the electronic partition functions that have been listed by Hansen (1976).

$$Q_{el}(H) = 2$$

$$Q_{el}(N) = 4 + 10\exp(-27,658 / T) + 6\exp(-41,495 / T)$$

$$Q_{el}(O) = 5 + 3\exp(-228.9 / T) + \exp(-325.9 / T) + \\ + 5\exp(-22,830 / T) + \exp(-48,621 / T)$$

$$Q_{el}(N_2) = 1$$

$$Q_{el}(O_2) = 3 + 2\exp(-11,393 / T) + \exp(-18,985 / T)$$

$$Q_{el}(NO) = 2 + 2\exp(-174.2 / T)$$

$$Q_{el}(H_2) = 1$$

$$Q_{el}(Cl_2) = 1 + 6\exp(-26,345 / T)$$

This Appendix is not intended to be a handbook for the best available data on the fullest possible range of gas species. It merely provides reasonable values, for some common gases, of the physical quantities that are required for the application of the DSMC method. As was illustrated in §8.3, the uncertainties in DSMC results that are associated with the molecular properties are far greater than the likely computational errors

The values of the physical constants are included in the List of Symbols.

References

Bird, G. A. (1994) *Molecular Gas Dynamics and the Direct Simulation of Gas Flows,* Clarendon Press, Oxford.

Chapman, S. and Cowling, T.G. (1970). *The Mathematical Theory of Non-uniform Gases* 3rd edn., Cambridge University Press.

Hansen, C. F. (1976) *Molecular physics of equilibrium gases, a handbook for engineers,* NASA SP-3096.

Park, C. (1990) *Nonequilibrium hypersonic aerothermodynamics,* Wiley, New York.

APPENDIX B. Downloadable files and links

While all the executable programs are for Windows, the Fortran source code can be compiled for any operating system that supports a modern Fortran compiler. Versions of the Xojo (Real Studio) program is available for Windows Linux and IOS. Moreover, each of these programs can produce executables for all three operating systems

B.1 Files associated with the DSMC program

Name	Date modified	Type	Size
DS1D Libs	26/07/2013 11:21 ...	File folder	
QKrates Libs	20/02/2013 10:46 ...	File folder	
DS1D.EXE	14/02/2013 6:54 PM	Application	2,858 KB
DS1D.rbp	14/02/2013 6:55 PM	Real Studio Project	440 KB
DSMC.f90	25/07/2013 10:48 ...	Fortran Source	195 KB
QKDATA.H2O2	4/01/2013 9:24 AM	H2O2 File	1 KB
QKdata.REAL_AIR	7/12/2012 2:13 PM	REAL_AIR File	1 KB
QKrates.exe	20/02/2013 10:46 ...	Application	3,309 KB
QKrates.rbp	20/02/2013 10:47 ...	Real Studio Project	1,034 KB

DSMC.F90   The free-form Fortran source code. Executables are not provided because the output files are often altered to suit particular applications. The source code contains the code to generate all these alternate file options, but most of the statements are commented out when not required. Intel Fortran Compiler XE 13.1 was used to compile this program. Any modern Fortran compiler should suffice, but without access to the extended mathematics library, it may be necessary to add subroutines for error and gamma functions.

DS1D.EXE   When executed, it produces an interactive menu for the generation of **DS1D.DAT** data files for applications of the DSMC code to one-dimensional flow applications.

DS1D.RBP   The RealStudio (now Xojo) source code that generates the executable.

DS1D.LIBS   A folder that contains dynamic link libraries that are required by **DS1D.EXE** and must be in the same folder as that file when it is run.

QKrates.EXE An interactive application for producing the data that is required in gas models that include chemical reactions. It is used to ensure that the data for the forward and reverse reactions satisfy the law of mass action. It also makes graphical comparisons with any experimentally based rate coefficients and the current implementation of the Q-K theory allows these, if required, to be brought into closer agreement. It is also used for the generation of new **QKrates.TXT** files. The application is discussed in §6.4 for dissociation-recombination and in §6.5 for exchange and chain reactions. It is almost essential for the development of reacting gas data.

QKrates.RBP The RealStudio (now Xojo) source code that generates the executable.

QKrates.LIBS A folder that contains dynamic link libraries that are required by **QKrates.EXE** and must be in the same folder as that file when it is run.

QKdata.REAL_AIR A file that contains the physical properties of all the molecular species that are involved in air at the temperatures that are associated with orbital re-entry speeds. It must be renamed to **QKdata.TXT** before **QKrates.EXE** is run if the real air reactions are to be studied.

QKdata.H2O2   A file that contains the physical properties of all the molecular species that are involved in the combustion of a hydrogen-oxygen mixture. It must be renamed to **QKdata.TXT** before **QKrates.EXE** is run if the real air reactions are to be studied.

The link that allows the downloading of the DSMCfiles.zip folder that contains these files is:

https://dl.dropboxusercontent.com/u/82909476/DSMCfiles.zip

B.2 Files associated with the DS2V program

Name	Date modified	Type	Size
Appearance Pak.dll	11/01/2012 5:37 PM	Application extens...	136 KB
DS2.exe	14/07/2013 4:13 PM	Application	1,798 KB
DS2.F90	14/07/2013 4:12 PM	Fortran Source	518 KB
DS2_32bit.exe	5/01/2012 7:20 PM	Application	1,437 KB
DS2V.exe	22/07/2011 2:45 PM	Application	14,249 KB
DS2V4.RBP	1/07/2011 6:35 PM	Real Studio Project	2,321 KB
DS2VD.DAT	11/01/2012 5:37 PM	DAT File	1 KB
DS2VDataSpec.txt	11/01/2012 5:37 PM	Text Document	15 KB
RBQT.dll	11/01/2012 5:37 PM	Application extens...	128 KB
Shell.dll	11/01/2012 5:37 PM	Application extens...	96 KB

DS2.F90   The free-form Fortran source code. Executables are provided because of the complexity of the source code is such that it should be altered or extended only with extreme caution. Intel Composer XE was used to compile this program. Any modern Fortran compiler should suffice because the code does not make use of nonstandard mathematical libraries.

DS2.exe   The 64 bit (x86) version of the compiled program can be run as a stand-alone console application or, preferably, interactively from within **DS2V.exe** as long as that application is in the same folder.

DS2_32bit.exe A 32 bit (Win32) version for running on older computers that do not support x64 executables. It must be renamed to **DS2.exe** if run within **DS2V.exe**.

DS2V.exe   An shell program for interactive running **DS2.exe**. It provides progressive contour plots of flowfield properties and line plots of surface properties as well as a comprehensive display of the information relating to the run. It also incorporates menus for producing the **DS2V.DAT** data programs. It can be run in a post-processing mode and has an option for generating movies from runs with continuing unsteady sampling.

AppearancePak.dll, RBQT.dll and Shell.dll
> Dynamic link library files that must be in the same folder as **DS2V.exe** when it is run.

DS2V4.RBP The RealStudio (now Xojo) source code that generates the executable.

DS2VD.DAT A sample data file for the demonstration case in §8.2.

DS2VDataSpec.TXT The specification of the **DS2VD.DAT** files.

The link that allows the downloading of the DS2Vfiles.zip folder that contains these files is:

https://dl.dropboxusercontent.com/u/82909476/DS2Vfiles.zip

Disclaimer

The purpose of the programs is to demonstrate the DSMC method. They are not guaranteed to be free from error and should not be relied upon for solving problems where an error could result in injury or loss. If the programs are used for such applications, it is entirely at the user's risk.

Results can be attributed to these programs only if there have been absolutely no changes to the source code.

APPENDIX C. Current DSMC issues

The demonstration cases provide an indication of the capabilities of the general purpose DSMC codes that have been developed by the author over the past twenty years and these were based on the special purpose programs that had been developed over the preceding thirty years. The objective has been to develop executable programs, or "apps" in the modern terminology, that can be readily applied to practical problems by non-specialists. To this end, the computational variables such as grid size and the magnitude of the time steps are calculated automatically within the program. Moreover, the programs are intended to be largely self-validating. This Appendix is concerned with the extent to which this has been achieved and with a discussion of some particular difficulties that have arisen.

Computer capabilities and the DSMC procedures have changed over the period and one issue is that it is still necessary to use older codes that do not take full advantage of these developments. More specifically, the **DSMC** code does not yet have two and three-dimensional capabilities and the two-dimensional cases in Chapter 8 were calculated with the now obsolete **DS2V** program. This has been applied successfully by hundreds of workers, but is no longer supported. The source code has therefore been made available so that problems can be investigated and extensions could be made. However, the history of the code is such that it is now almost unsupportable. For example, there have been two major changes to the geometry model and there has also been a change in the application that has been used for the generation of the graphical user interface, or GUI. Any change to the source code can have unintended consequences and, even if the user is sufficiently experienced to realize that something has gone wrong, the rectification can be very time consuming. Any attempt to use the current code as the basis of an improved or extended code is strongly discouraged.

There are no three-dimensional cases in this version of the book. The **DS3V** program employed the earlier application for the generation of the graphical user interface. This provided Fortran subroutines that accessed the Windows application program interface, or API. At least one of these DS3V routines no longer works with Windows 7 and

8 and the program is unusable on the latest PC's. In addition, the 32-bit round-off errors that are only occasionally a serious problem in **DS2V** had, in any case, made **DS3V** almost unusable.

A problem with the separate programs **DS1V**, **DS2V** and **DS3V** for one, two and three-dimensional flow classes was that, if any change was made to the collision subroutine that includes the very complex procedures for chemical reactions, similar changes had to me made to all programs. The **DSMC** program will cover all flow classes of flow and the collision subroutine that has already been developed for zero and one-dimensional flows will be used for all flows.

A further problem is that the earlier programs were not parallel programs and parallelism must be employed if full advantage is to be taken of modern CPU's. It is planned that, once the **DSMC** code is able to treat all classes of flow, it will be made into a parallel code. It will distribute the program loops between the processor cores rather than employ the domain decomposition that is now the basis of most current parallel codes.   This assumes that the purpose of the parallelism is to reduce the duration of the run. There is a simpler way of taking advantage of multiple core CPU's if the objective is, instead, to just increase the size of the sample. As discussed in §1.4, multiple instances of the same program can be run concurrently with the output being based on the ensemble average. This facility will also be added to later versions of the **DSMC** code.

The strength of a direct physical simulation, as opposed to the numerical solution of mathematical equations, is that numerical instabilities can be expected to be completely absent. However, there have been two instances where non-physical effects have been observed in DSMC calculations.

The first is a spike in the heat transfer at the centre-line of an axially symmetric blunt body in hypersonic flow. This has appeared in a number of calculations and appears to be more severe when nearest-neighbour collisions are employed. The spurious nature of the spike is obvious when it is about 10-20% of the incident flux, but can be missed when it is only a few percent. For example, the heat transfer distribution for the re-entry blunt-body demonstration case in Fig. 8.30 does not exhibit an obvious anomaly. However, this is a logarithmic plot and the linear plot of heat transfer as a function of distance from the stagnation point in Fig. C.1 is anomalous in that the slope should be zero at the stagnation point.

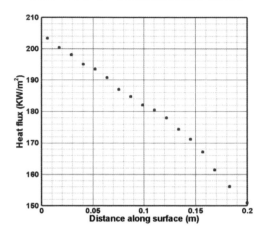

Fig. C.1 The heat transfer in the §8.5 demonstration case.

The result in Fig. C.1 is based on a long time average and averages over shorter time intervals generally led to values below 200 kW/m², but with occasional excursions well above this value.   It therefore appears that there is a small spike in this case.   The spike can be a much larger fraction of the net flux when (as is often the case with this class of flow) the net heat flux is the small difference between the incident and reflected fluxes.   In fact, at lower altitudes where DSMC overlaps with continuum CFD, the net heat flux is a very small difference and it is almost essential to obtain it indirectly from the temperature gradient at the surface.

The basic problem appears to be with the almost unavoidable use of radial weighting factors for this class of flow.   While these are correct on the average, the departures from the average due to statistical scatter give rise to random walks in the flow densities.   If these fluctuations cause a sufficiently large departure over a sufficiently large area, sonic waves are produced that are enormously amplified if they propagate to the axis.   Support for this can be obtained by running the program in an unsteady sampling mode because the spikes at the axis are then, as described in the preceding paragraph, seen as occasional effects. The only solution would appear to be to make two identical but statistically independent calculations and to periodically swap the simulated molecules in a "checkerboard" fashion in order to break up the spurious waves. Should this theory be found to be correct, the DSMC program will be modified to implement the fix.

While the first problem appears to have a physical origin, the second is definitely numerical. During test calculations of shock wave structure with the **DS1V** program, unpredictable and clearly unphysical bumps appeared on one occasion in the upstream region of the wave. The **DSMC** program is based on **DS1V** and the shock wave profiles in §7.3 are just a few of the many that have since been calculated. The bumps have not reappeared and, because it was thought that the time step may have been a factor, the Mach 2 case has been recalculated with a progressive doubling of the time steps. A uniform thickening of the wave became noticeable when the time step reached sixteen times the automatically set step, but no bumps were produced. The effect is evidently rare and might even have been caused by a bug in the code that has since been fixed. However, it remains a concern because it was the only occasion during 50 years of DSMC calculations by the author that an instability of the type that plagues continuum CFD has appeared.

It may be noted that the flows in all the twelve examples in Chapter 8 involve vortices, albeit undesirable vortices in §8.4 and §8.13. A modern application with millions of collision cells that each contain the optimum eight simulated molecules has meaningful flow gradients within a sampling cell and these may embody strong vorticity. The smallest vortices appear to have diameters of ten to fifteen mean free paths and, as long as the mcs/mfp ratio is small compared with unity, the vorticity is effectively preserved in collisions. There had been claims (e.g. Meiburg, 1986) that DSMC cannot adequately model flows with vorticity, but these were based on the misguided assumption that DSMC necessarily involves mcs/mfp ratios that are large in comparison with unity.

Validations against experiment have shown that DSMC is equally successful for flows with and without vortices. For example, Moss and Bird (2004) compared the results of **DS2V** (Version 3) calculations of the Mach 15.6 flow of nitrogen over a $25°/55°$ biconic with measurements in the 48-inch shock tunnel at CUBRC, Buffalo NY. The Knudsen number based on the diameter is $8×10^{-4}$. Figure C.2 shows the temperature and mcs/mfp contours for a calculation with eight million simulated molecules. The flow separates upstream of the junction of the cones and reattaches upstream of this junction. The key comparison with experiment is for the separation and re-attachment points that are indicated by **S** and **R** in Fig. C.2.

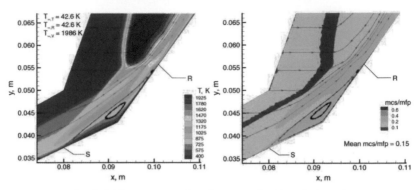

Fig. C.2 The separated region near the biconic junction.

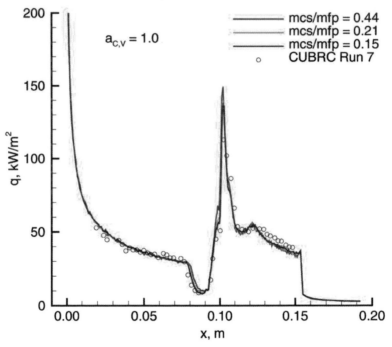

Fig. C.3 Calculated and measured heat transfer distribution over the biconic.

The separation point is marked by a sharp dip in the heat transfer and the re-attachment point by kink in the profile. The extent of the separation increases as the mcs/mfp radio is reduced through increases in the total number of simulated molecules. It appears that the converged value is very close to the measured value.

Table C.4 Effect of the mean mcs/mfp on the extent of the separation region.

Program	Total molecule number	Mean mcs/mfp	Extent of separation (mm)
DS2V v.3	500,000	0.94	17.0
DS2V v.3	2,000,000	0.44	19.3
DS2V v.3	8,000,000	0.21	20.4
DS2V v.3	16,000,000	0.15	20.9
SMILE	79,400,000	-	21.5

Further details of the convergence of the separation extent with decreasing mcs/mfp are provided in Table C.1. An 80 million molecule parallel calculation with the SMILE program was in similar agreement with the measured profile and yielded a slightly more converged result. The DS2V calculations were made with Version 3 which did not employ nearest-neighbour collisions. A similar calculation with Version 4 would be expected to at least equal the SMILE result.

With new capabilities for two and three-dimensional flows being developed as part of the DSMC program, the question arises as to whether these can provide a convergence behaviour that is significantly better than the DS2V program. The largest potential gain is through improving the quality of the cells near surfaces. The current procedures grow cells by adding the elements nearest the cell node with no control over the cell shape. Experience with other cell schemes has shown that much is to be gained through elongated cells that are parallel to any adjacent surface. If the elements were to be assigned levels relative to the surface, as is already done with the divisions (see Fig. 4.6), the desired "body-fitted" cells could be grown.

References

Meiburg, E. (1986). Comparison of the Molecular Dynamics method and the direct simulation for flows around simple geometries. *Phys. Fluids* **29**, 3107-3113.

Moss, J. N. and Bird, G. A. (2004). DSMC Simulation of Hypersonic Flows with Shock Interactions and Validations with Experiments. Paper presented at the 37th AIAA Thermophysics Conference, Portland OR.

298

SUBJECT INDEX

Definitions are denoted by bold type.

Printed in Great Britain
by Amazon